# PROBLEM SOLVING IN GCSE MATHEMATICS

*Questions for pupils targeting Grades 8 and 9*

Daniel Griller

Problem Solving in GCSE Mathematics



Cover design by Dex

ISBN 978-0-9957826-1-7

## Preface

The aim of this book is to provide GCSE and IGCSE Mathematics students with a self-contained source of challenging problems as they prepare for their exams.

The questions have been arranged in approximate order of difficulty; everyone has different strengths and weaknesses, but I would expect most pupils to find the first few problems relatively straightforward (compared to those appearing later on), whereas the last few problems are significantly tougher, and would almost certainly give even the most able candidates pause for thought.

Problems for which calculators are not permitted, or that belong to only one of the two qualifications, are indicated as such in the text.

The final publication is by no means a solo effort. I am hugely grateful to my proof-readers Marta Watson-Evans, Ashley Page and Rosamund Bradbury, and special thanks must go to Tom and Naomi Bowler, who dedicated a significant number of hours to working diligently through the entire text, providing valuable feedback on both the English and mathematical components of the problems and solutions.

My greatest hope is that you find these problems useful, and that they help you achieve an excellent final grade. That being said, there is nothing quite like the satisfaction of solving a difficult problem, and you might just find yourself enjoying the task. Which is fine by me.

*Daniel Griller, January 2019*

# Table of Contents

# PROBLEMS

1. The mass of a ball bearing is measured as 10.2 g to the nearest tenth of a gram, and its volume as 2.54 $cm^3$ to three significant figures. Determine, to two decimal places, upper and lower bounds for the density of the ball bearing, in grams per cubic centimetre.

---

2. *(Non-calculator)* Evaluate the following numbers, leaving your answers in standard form:

(a) $\left(1.7 \times 10^{200}\right) - \left(1.8 \times 10^{199}\right)$

(b) $\left(4.3 \times 10^{-55}\right) \div \left(5 \times 10^{-122}\right)$

3. *(Non-calculator)* Estimate the value of $\sqrt[3]{\frac{11.8\times30.1}{2.94}}$, giving your answer to one significant figure.

---

4. Make $q$ the subject of the formula $p=\frac{3+\sqrt{q}}{3-\sqrt{q}}$

5. *(Non-calculator)* Express $0.47\dot{6}$ as a fraction in its lowest terms.

---

6. Isabella is picking an outfit for a mild winter evening. She needs to choose between 3 jumpers, 4 pairs of trousers and 4 pairs of boots. Additionally, she might wear a scarf (3 to choose from) and she might wear a hat (2 to choose from). In how many different ways can she select an outfit?

7. (a) Mayu says

*"If the product of two numbers is an integer, then at least one of the numbers must be an integer."*

Is Mayu correct? Explain your answer.

(b) Naoki says

*"If two lines are perpendicular, then the product of their gradients is always equal to -1."*

Is Naoki correct? Explain your answer.

8. The following table and histogram display the total number of coffee cups sold in a single day by each of 100 different cafés in a big city.

| Number of cups sold $(x)$ | Frequency |
|---|---|
| $0 \le x < 100$ | |
| $100 \le x < 150$ | 15 |
| $150 \le x < 200$ | 22 |
| $200 \le x < 250$ | 27 |
| $250 \le x < 300$ | 19 |
| $300 \le x < 500$ | |

(a) Use the histogram to complete the table.

(b) Use the table to complete the histogram.

Frequency density

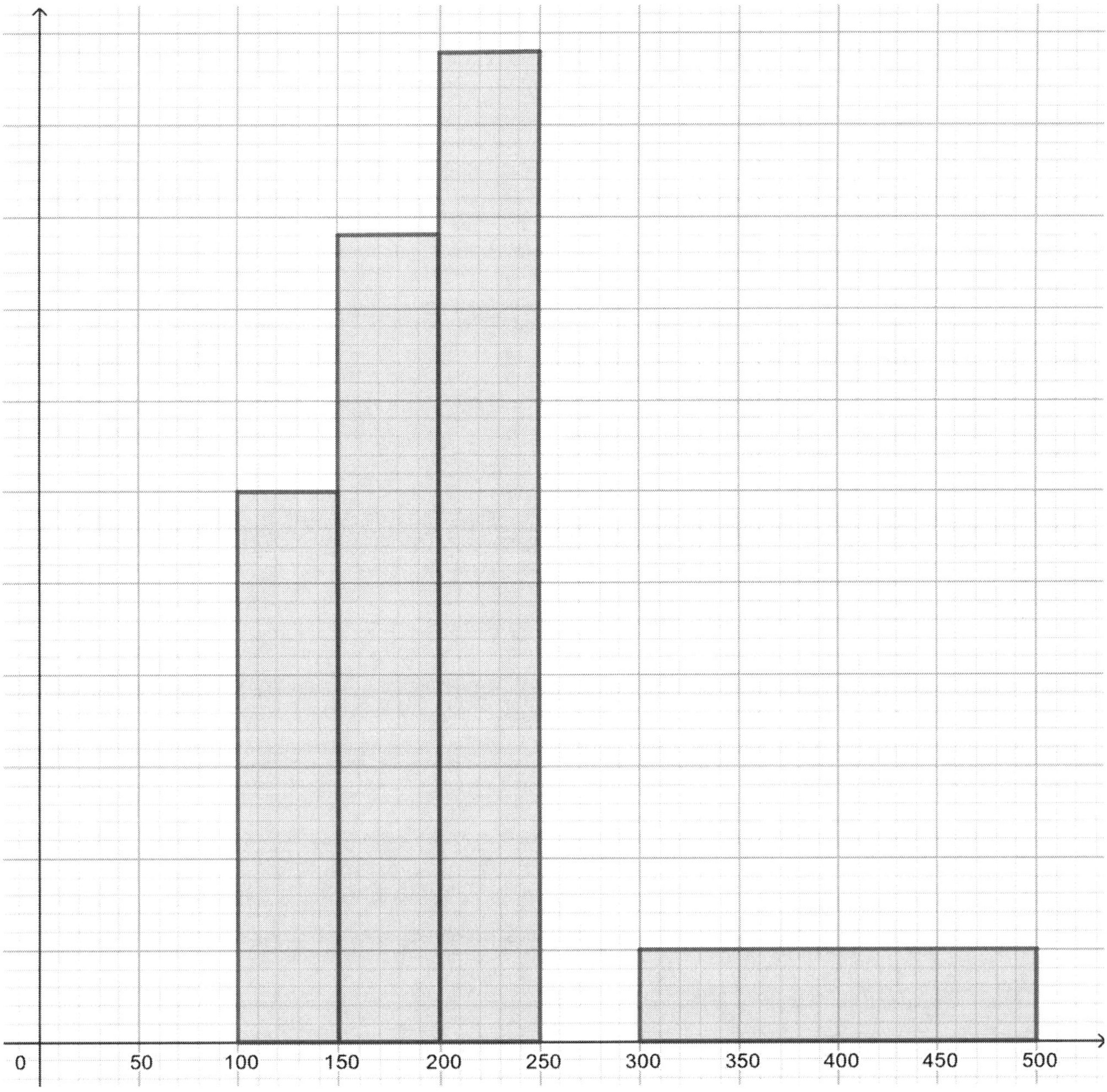


Number of coffee cups sold

9. Expand and simplify $(y-2)(2y-3)(y^2+4)$

10. Solve the following simultaneous equations graphically, by drawing suitable lines on the axes below.

$$x + 2y = 3$$

$$2x - y = -4$$

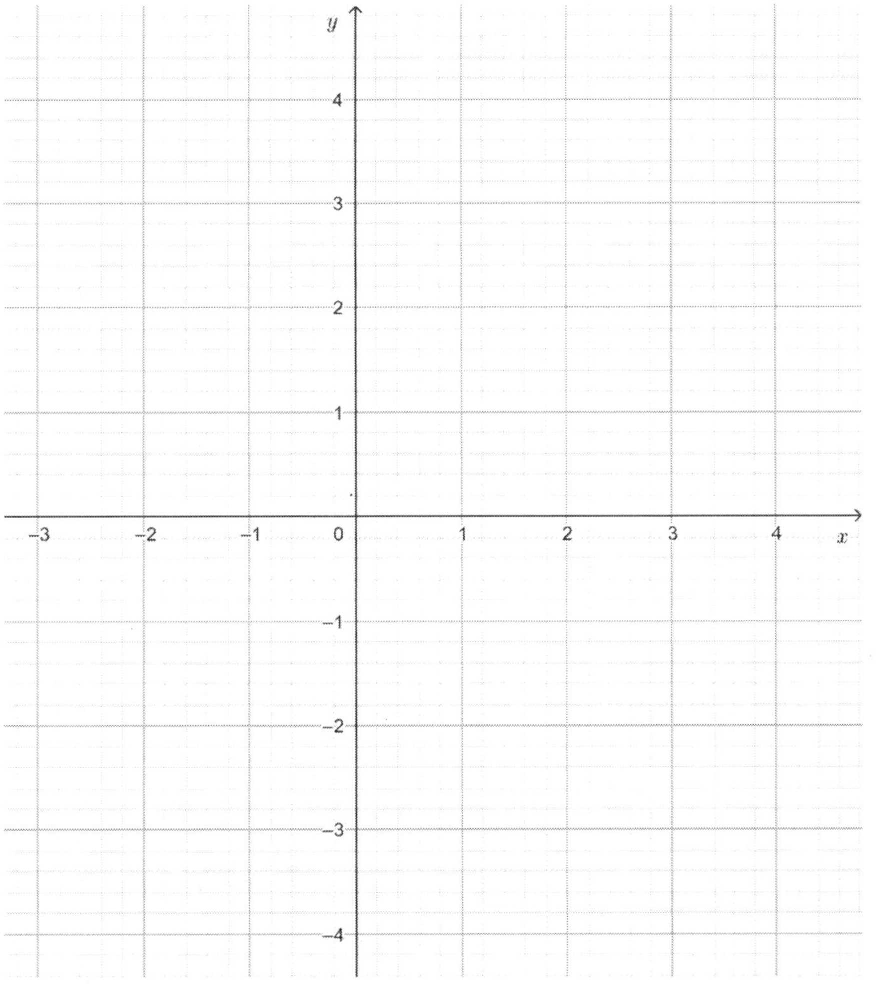

11. Factorise fully the expression $75ab^2 - 12a^3$

---

12. *(Non-calculator)* Alice is thinking of a positive integer $N$. The lowest common multiple of $N$ and 45 is 90. Find all possible values of $N$.

13. Paradise Money allows holidaymakers to exchange pounds (£) for dollars ($) and vice versa.

They sell dollars at a rate of $1.27 per £ (rounded to the nearest $0.01).

They buy dollars at a rate of $1.37 per £ (rounded to the nearest £0.01).

Galip purchases £250-worth of dollars in notes and coins at Paradise Money for a trip to the United States. When he is there, he ends up using his credit card for everything, without spending any of his American cash. When he returns to the UK, he goes back to Paradise Money to convert all of his dollars back to pounds.

How much has he lost, in pounds and pence, as a result of the two exchanges?

14. A force $F$ acts on a particle $P$, so that $F$ is inversely proportional to the cube root of the distance between $P$ and a fixed point $O$. When $OP = 8\ \text{m}$, $F = 15\ \text{N}$

Calculate the distance $OP$ at the moment this distance (in metres) is numerically equal to the force $F$ (in Newtons), giving your answer to three significant figures.

---

15. *(IGCSE)* Calculate the difference between the sum of the first 100 positive multiples of 7 and the sum of the first 100 positive multiples of 9.

16. Martina lives and works in London, but she is due to be at a breakfast business meeting on Wednesday 22nd August in New York City, at 8am local time.

Flights from London to New York are 8 hours long and depart every hour on the hour.

Martina has calculated that it will take her 45 minutes to travel from her home to the airport, where she plans to arrive 2 hours before departure time. Once she lands in New York, she will need 2 hours to prepare for and arrive at the breakfast meeting. She can sleep on the plane.

Given that the time difference between New York and London is 5 hours (New York City is 5 hours behind London), what is the latest date and time she can leave her house in order to get to the meeting on time?

17. *(IGCSE)* In the diagram below, $A, B$ and $C$ are sets.

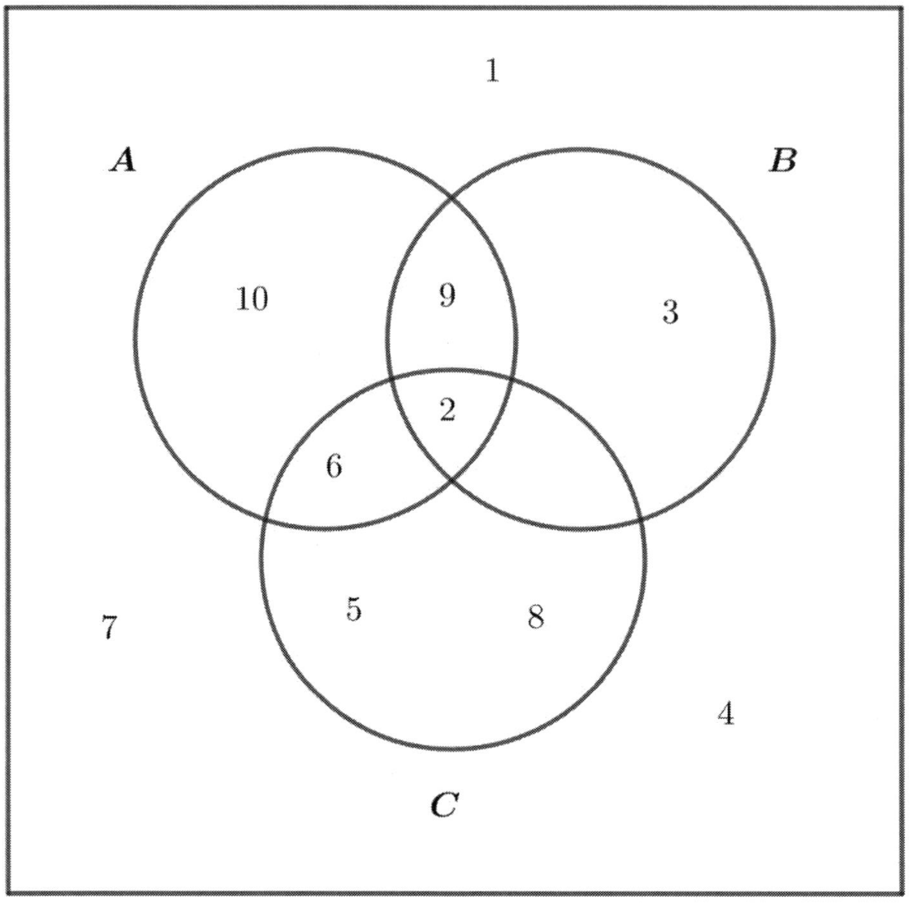


List the elements that appear in:

(i) $A \cup C'$

(ii) $(A' \cap B) \cup C$

18. *(Non-calculator)* Between August and September, the price of a T-shirt fell by 40%

Between September and October, the price fell by 10%

The difference between the August and October prices was £20.70

Calculate the price of the T-shirt in September.

19. On a farm, there are cows, sheep and pigs, with no other animals. For some number $n$,

the ratio of cows to sheep is $n : n + 7$

the ratio of sheep to pigs is $n : n + 10$

and the ratio of cows to pigs is $n : n + 19$

What fraction of all the animals are sheep?

20. *(Non-calculator)* Suppose

$$x + \frac{1}{y} = 3.125$$

Find the decimal equal to

$$\frac{y}{xy + 1}$$

---

21. In a geometric sequence of positive numbers, the first term is 5, and the third term is 100.

Calculate the sum of the first five terms of the sequence. Leave your answer in the form $a + b\sqrt{5}$, where $a$ and $b$ are integers.

22. *(Non-calculator)* Solve

$$\frac{38}{43 - x^4} = \frac{43}{38 - x^4}$$

---

23. Prove or disprove the following claims:

(i) The sum of any four consecutive multiples of 5 is divisible by 20

(ii) The sum of any five consecutive multiples of 4 is divisible by 20

24. The two sculptures below are mathematically similar.

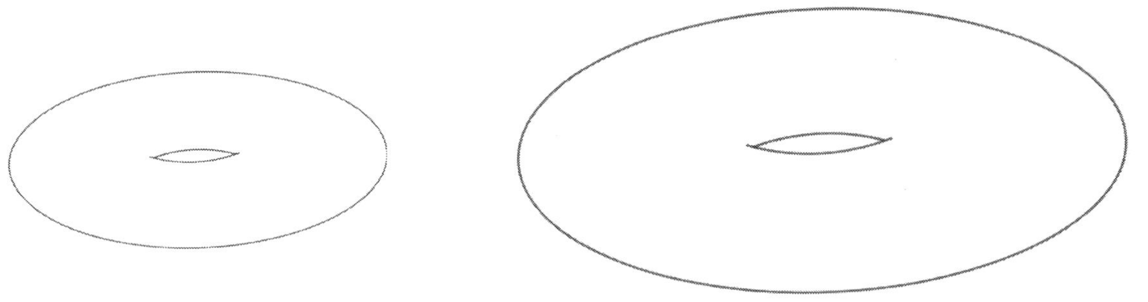

The larger sculpture has surface area 240 cm$^2$ and volume 80 cm$^3$

The smaller sculpture has volume 45 cm$^3$

Calculate its surface area, to three significant figures.

25. *(Non-calculator)* A circle centred at the origin passes through the point $(-3,5)$

Find an equation for

(a) the circle

(b) the tangent to the circle at $(-3,5)$, giving your answer in the form $y = mx + c$

26.

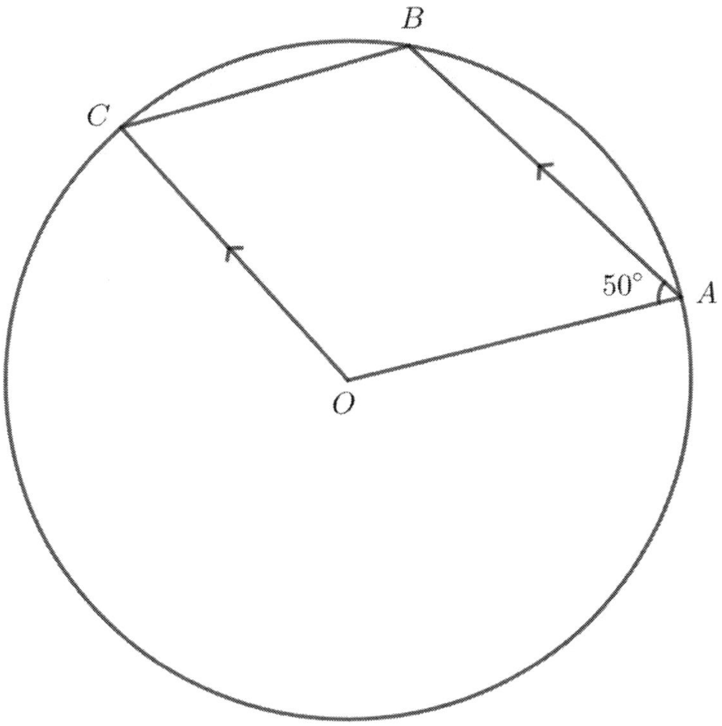


In the diagram above, $O$ is the centre of the circle, $OC$ is parallel to $AB$ and angle $OAB = 50°$

Calculate the size of angle $OCB$

27. Write the correct equation underneath each graph. Choose from the list of equations below.

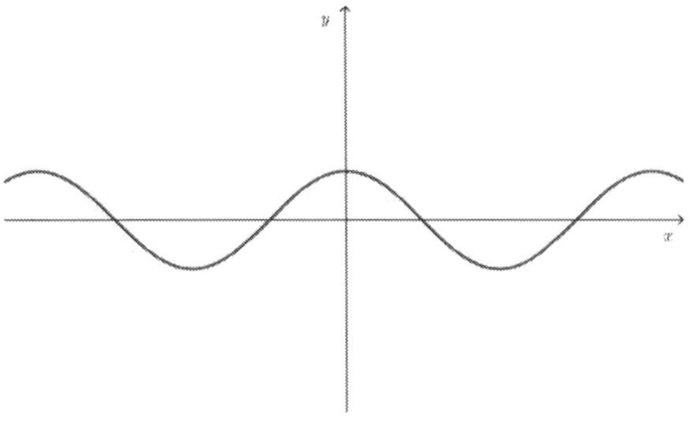

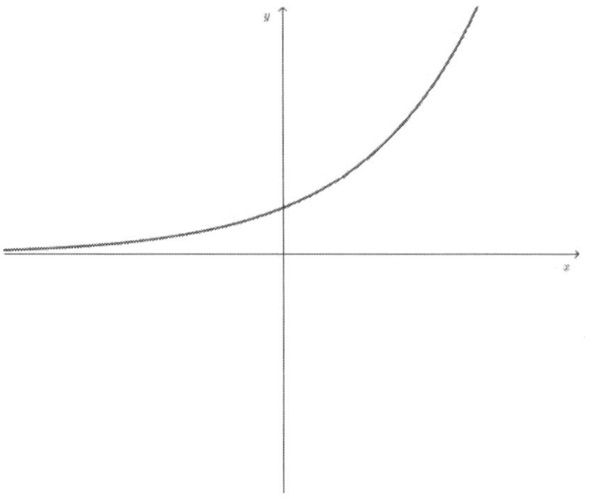

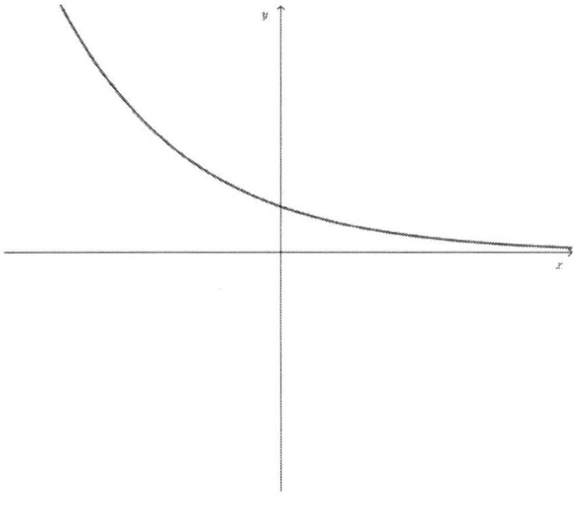

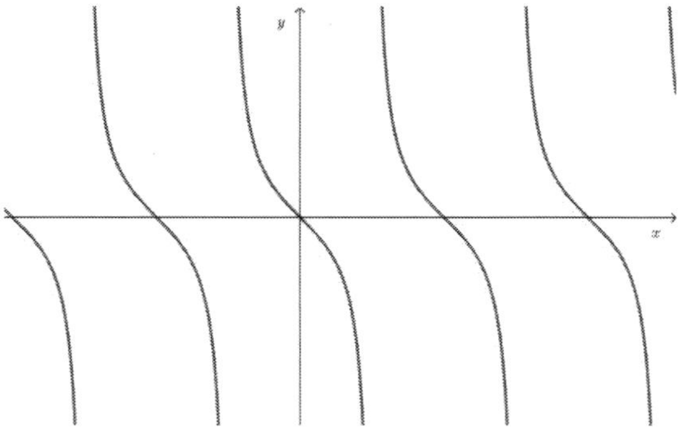

$$y = \left(\frac{3}{2}\right)^x \qquad y = -\left(\frac{3}{2}\right)^x$$

$$y = \sin(x) \qquad y = -\sin(x)$$

$$y = x^{-1} \qquad y = -x^{-1}$$

$$y = \cos(x) \qquad y = -\cos(x)$$

$$y = 0.7^x \qquad y = -0.7^x$$

$$y = \tan(x) \qquad y = -\tan(x)$$

28. *(Non-calculator)* A family of four is sharing a pizza for dinner. Mum takes $\frac{2}{5}$ of the pizza, Dad takes $\frac{3}{5}$ of the remaining pizza, Giulia splits what is left in the ratio 8 : 7 and takes the bigger slice for herself, leaving Matej with the smaller slice.

What percentage of the pizza does Matej receive?

---

29. *(Non calculator)* Lucas thinks of a positive number. He divides 15 by this number, and ends up with 1 more than 6 times his original number. What was the number he first thought of?

30. In the quarter circle below, the area of the segment is 1 cm$^2$

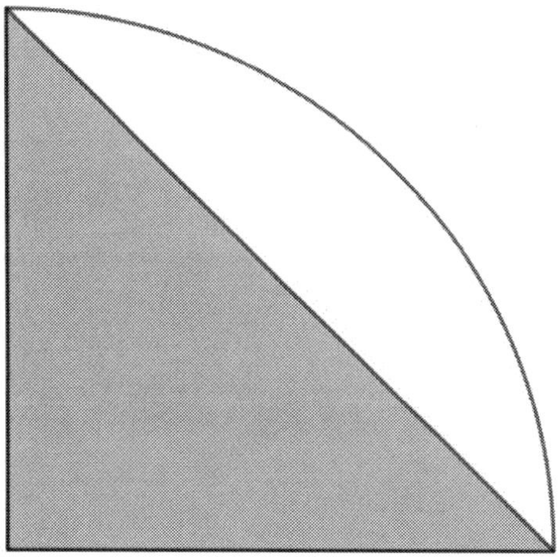

What is the exact area of the shaded triangle? Give your answer in terms of $\pi$

31. Below is the curve with equation $y = x^2 - 2x - 2$

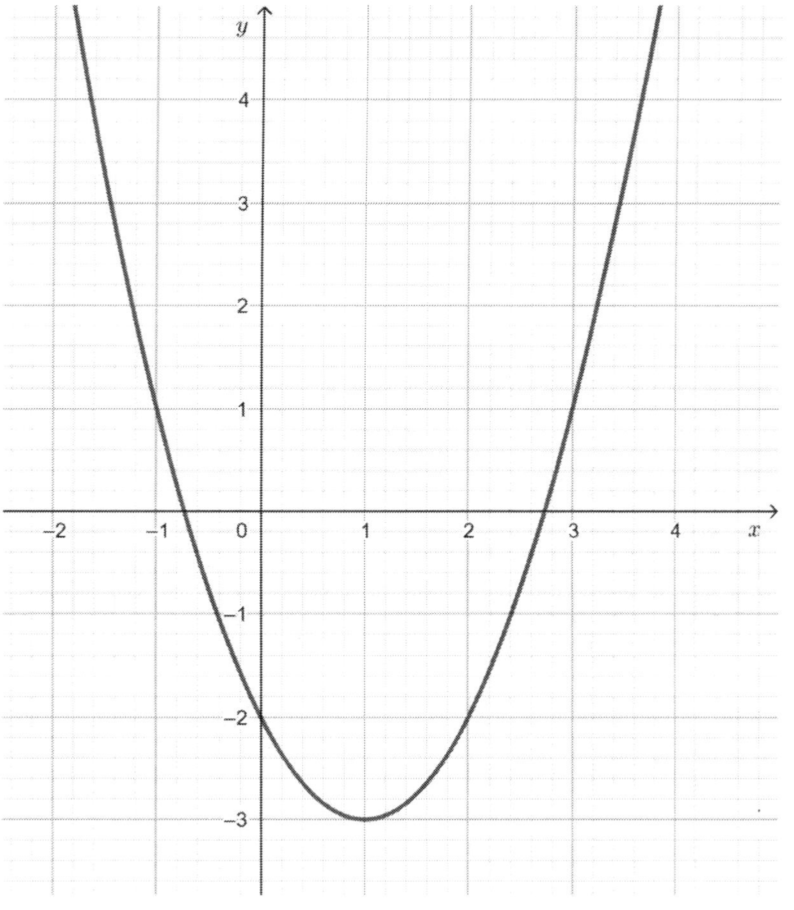


By drawing a suitable straight line on the diagram, estimate the solutions to

$$x + \frac{2}{x} = 3 + \frac{4}{x}$$

giving your answers to 1 decimal place.

32. A biased coin is more likely to land on Heads than on Tails. When it is flipped twice, the probability of obtaining one Heads and one Tails is 0.42

What is the probability of obtaining two Heads?

33. A sequence of hammers is drawn using dots. The first three hammers are shown below.

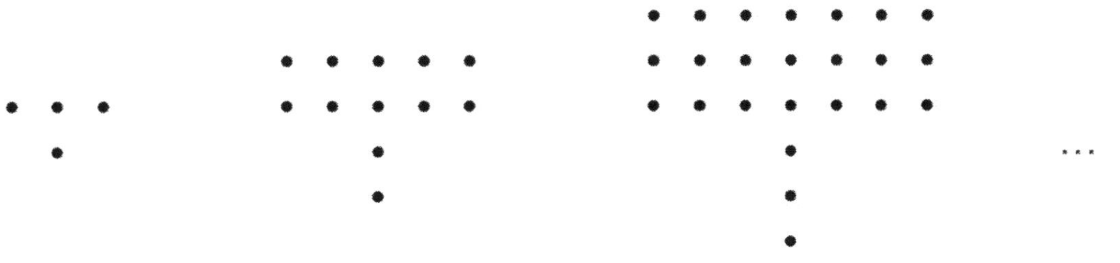

For every new hammer, an extra layer of dots is added to three sides of the hammer's head, and an extra dot is added to the handle.

Find a fully factorised expression for the number of dots required to draw the $n$th hammer, and explain why your expression is correct.

34. *(Non-calculator)* Malcolm either drives or cycles to work each day. On any given day, there is a 30% chance of rain.

If it rains, Malcolm is 5 times more likely to drive than cycle.

Otherwise, he is twice as likely to cycle as he is to drive.

(a) Complete the following probability diagram by writing fractions along the branches.

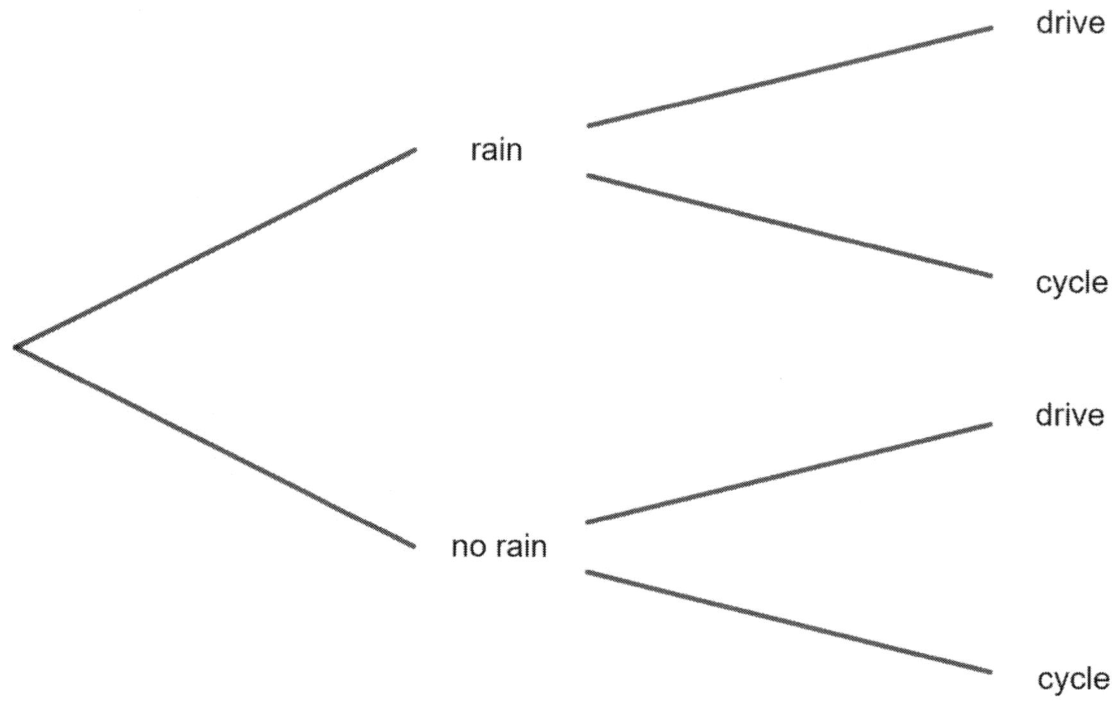

(b) Over the course of a year, Malcolm drove to work 145 times. On how many of these days would you expect there to have been rain?

---

35. Write

$$\frac{x-9}{x^2-9}-\frac{x-13}{x^2-2x-15}$$

as a single simplified algebraic fraction.

36. Find, to three significant figures, the shortest distance from the point (1,2) to the line $3x + 4y = 5$

37. Armani walks on a bearing of $100°$ for 4 km, then turns and walks on a bearing of $200°$ for 5 km. If she wishes to return directly to her starting point, on what bearing should she walk, and for what distance? Give your answers to 3 significant figures.

38. Define the functions $f$ and $g$ by

$$f(x) = \frac{1}{x-2}, \quad x \neq 2$$

$$g(x) = x^2 + 4x, \quad x \in \mathbb{R}$$

(a) Evaluate $f(1.8)$

(b) Find an expression for $g(2x-1)$ in the form $ax^2 + bx + c$

(c) Solve $gf(x) = 0$

39. *(Non-calculator)* Three square tiles of side lengths 5, 6 and 10 are placed on a square piece of paper as shown.

The difference in area between the two uncovered rectangular regions is $17cm^2$

Calculate the side length of the piece of paper, leaving your answer as a recurring decimal.

40. Here is a graph showing a skydiver's speed during the first 30 seconds after jumping from a plane.

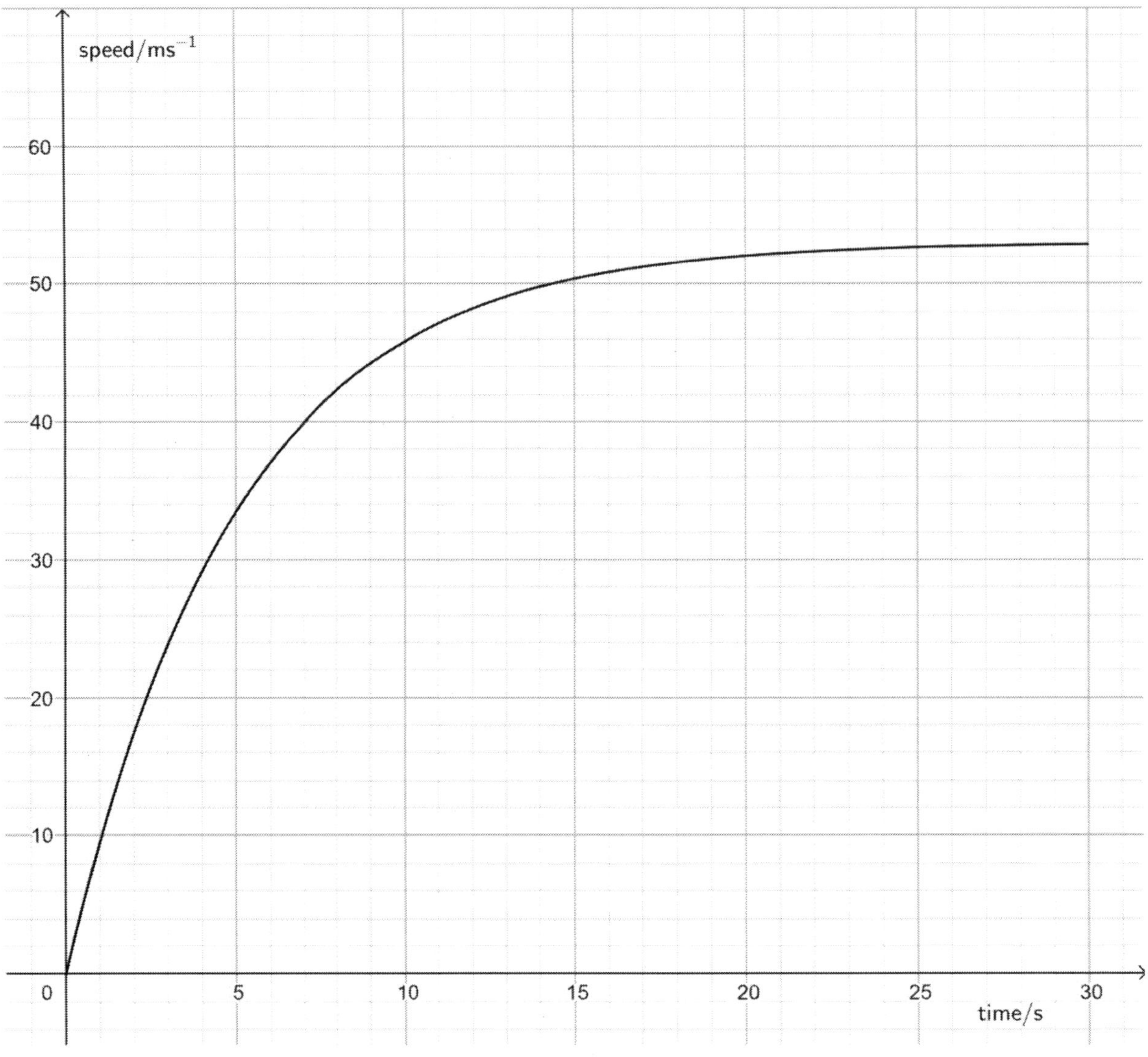


Use the graph to estimate, to two significant figures:

(a) the speed of the skydiver 5 seconds after the jump,

(b) the skydiver's acceleration 10 seconds after the jump,

(c) the distance travelled in the first 30 seconds after the jump (using four strips of equal width).

41. Triangle **P** is transformed into triangle **Q** by an enlargement, scale factor 2, centre $(-4,4)$

Triangle **Q** is then transformed into triangle **R** by a rotation of $180°$ about $(2,1)$

Describe fully the single transformation that will take triangle **R** back to triangle **P**

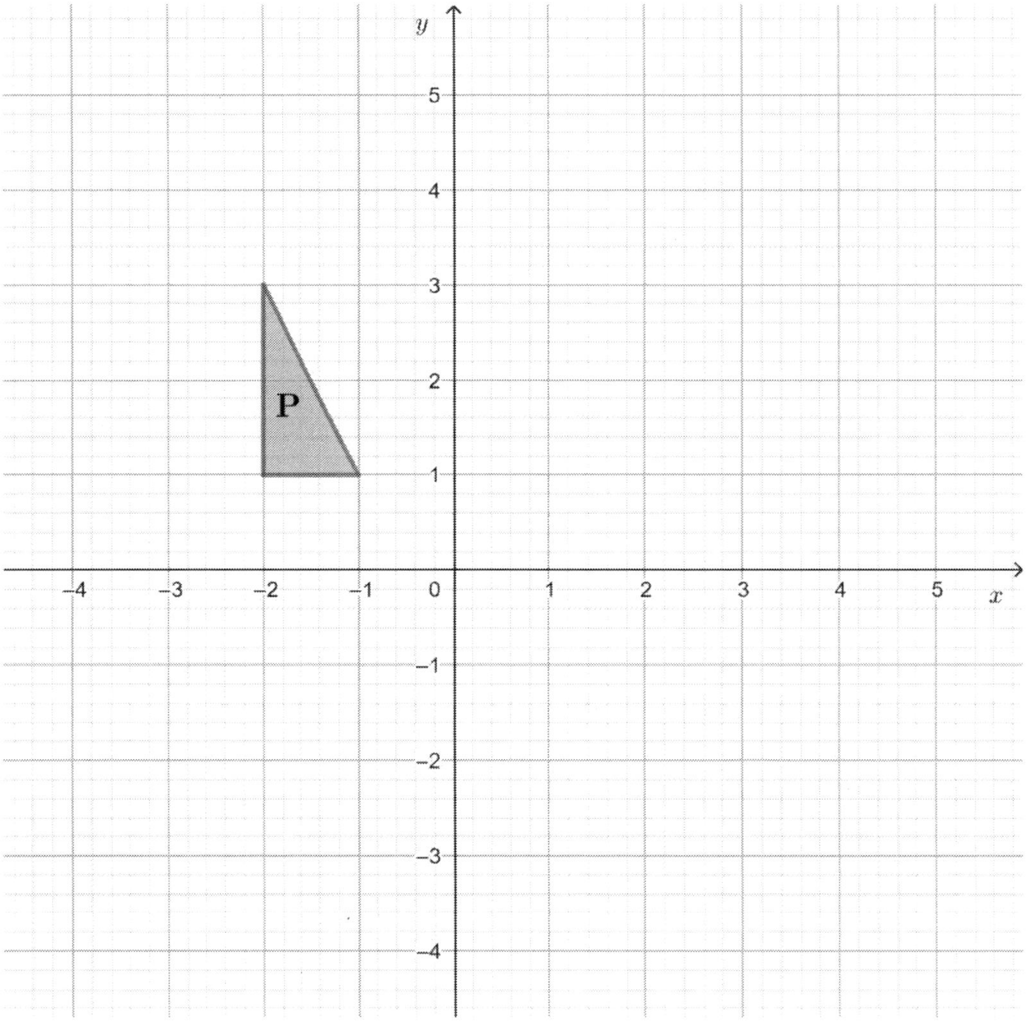

42. *(Non-calculator)* Solve simultaneously:

$$2x + 3y = -2$$

$$x(8 - 3y) = 28$$

43. Calculate the area of the triangle whose vertices have coordinates (1,5), (5,7) and (9,2)

44. Using the line below, construct a point $C$ so that angle $ABC = 67.5°$

Show all of your construction markings.

45. The number of cells in a jar is to be modelled by the equation

$$N = ar^t$$

where $N$ is the number of cells $t$ minutes after midday, and $a$ and $r$ are positive constants.

At 12:01pm, there are 2400 cells in the jar, and at 12:03pm there are 5400 cells.

Using the model given, predict the number of cells in the jar at 12:10pm, to the nearest hundred cells.

46. *(Non-calculator)* Find the value of $n$ for which $\sqrt[5]{2^n} = \sqrt{4^{n+1}}$

47. In the diagram below, $OA$ and $OB$ are tangents to a circle at $A$ and $B$ respectively.

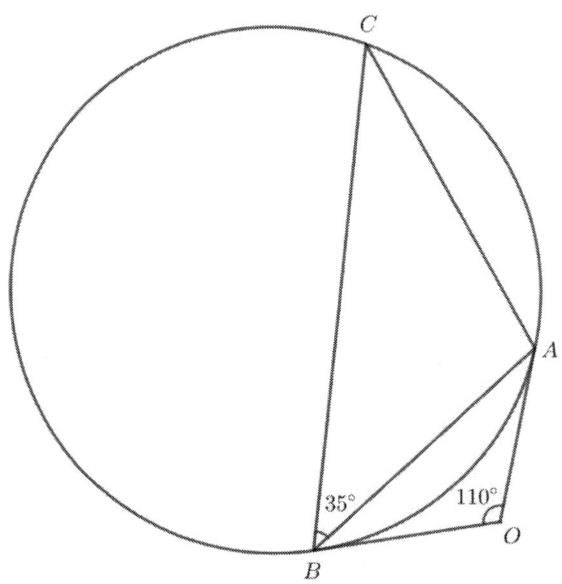

Prove that

(a) $OA$ is parallel to $BC$

(b) $OA \times BC = AB^2$

48. *(Non-calculator)* Express $\frac{\sqrt{8}}{4}+\frac{6}{\sqrt{32}}$ in the form $\frac{a}{b}\sqrt{c}$ , where $\frac{a}{b}$ is a simplified fraction and $\sqrt{c}$ is a simplified surd.

---

49. Treasure has been buried somewhere in the coordinate plane. You don't have a map, but you have some clues:

- The treasure is buried at a point $P$, somewhere on the straight line through (3,1) and (9,10)
- The distance from $P$ to (3,1) is twice the distance from $P$ to (9,10)

Find all possible coordinates of the point $P$

50. *(Non-calculator)* Elijah constructs a sphere and a cone out of solid gold, so that the base radius of the cone is equal to the radius of the sphere, the surface area of the cone is $k$ times the surface area of the sphere, and their volumes are equal.

Determine the exact value of $k$

51. Here is the graph of $y = f(x)$

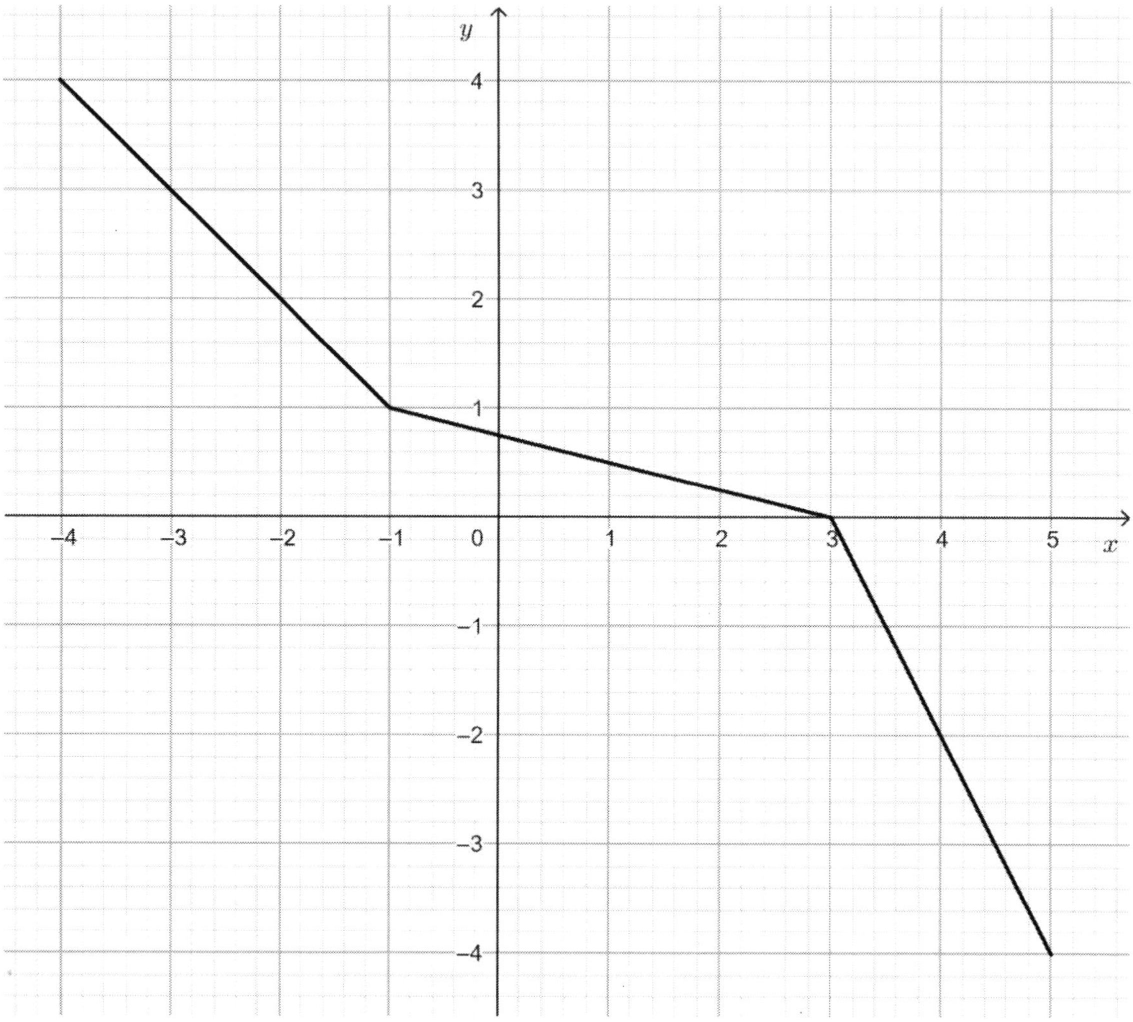


Write down

(a) The domain of $f$

(b) The range of $f$

Use the graph to evaluate the following:

(c) $f(4)$

(d) $ff(-3)$

(e) $f^{-1}(0.4)$

52. *(Non-calculator)* Evaluate the following, leaving your answers as integers or fractions in their simplest form:

(a) $9^{\frac{1}{3}} \div 9^{\frac{5}{6}}$

(b) $\left(\sqrt[37]{5}\right)^{111}$

(c) $16^{1.75}$

53. Let $f(x) = x^2 + 8x + 23$, where $x \in \mathbb{R}$

(a) Find an expression for $f(x)$ in the form $(x + a)^2 + b$

(b) Hence write down the coordinates of the turning point on the graph $y = f(x)$

Suppose we now restrict the domain of $f$ to be $x \geq -1$

(c) Find an expression for $f^{-1}(x)$

(d) State the domain of $f^{-1}$

54. Solve the inequality

$$(x + 1)(x + 2) < 12$$

and represent your answer on a number line.

---

55. *(GCSE)* We are looking for a positive solution to the equation

$$\left(x + \sqrt[4]{x}\right)\left(x - \sqrt[4]{x}\right) = 2$$

(a) Show that this equation can be rearranged to

$$x = \sqrt{2 + \sqrt{x}}$$

(b) Using the iterative formula

$$x_{n+1} = \sqrt{2 + \sqrt{x_n}}$$

with the starting value $x_0 = 1$, calculate values for $x_1$, $x_2$ and $x_3$ to 4 decimal places.

(c) Write down a solution of

$$\left(x + \sqrt[4]{x}\right)\left(x - \sqrt[4]{x}\right) = 2$$

giving your answer to 3 significant figures, and prove that your answer is correct to this degree of accuracy.

56. Calculate the area of the region defined by the following inequalities, to three significant figures.

$$1 + 2x > 0$$

$$3 - 4x > 0$$

$$x + y - 5 < 0$$

$$6x + 7y - 8 > 0$$

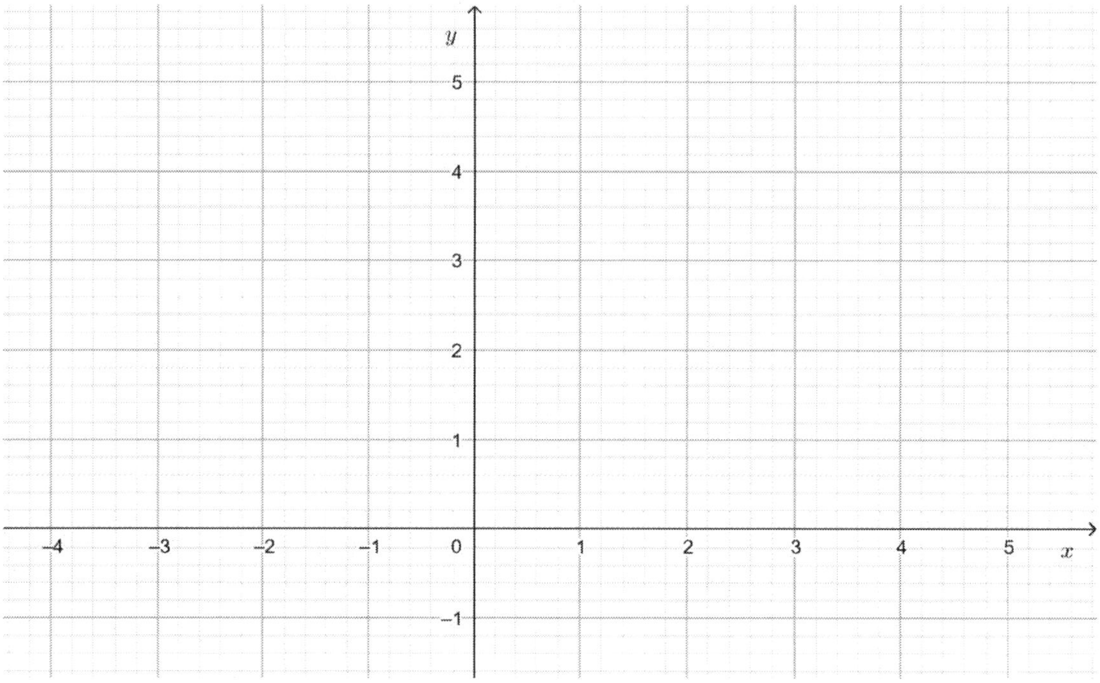

57. *(IGCSE)* Find the exact coordinates of all stationary points on the curve

$$y = \frac{(x+1)(x+2)}{3x}$$

and determine their nature.

58. Below is a graph of the function $y = f(x)$

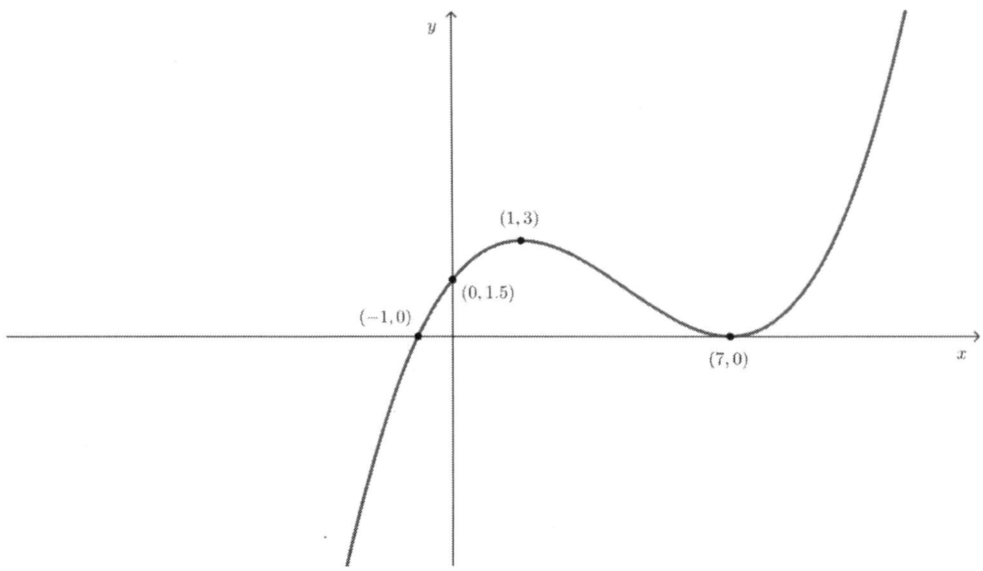


It meets the $x$-axis at $(-1,0)$ and $(7,0)$, crosses the $y$-axis at $(0,1.5)$ and it has a turning point at $(1,3)$

(a) On the axes below, sketch the graph of $y = f(1 - x)$, labelling all the points where the graph meets the axes.

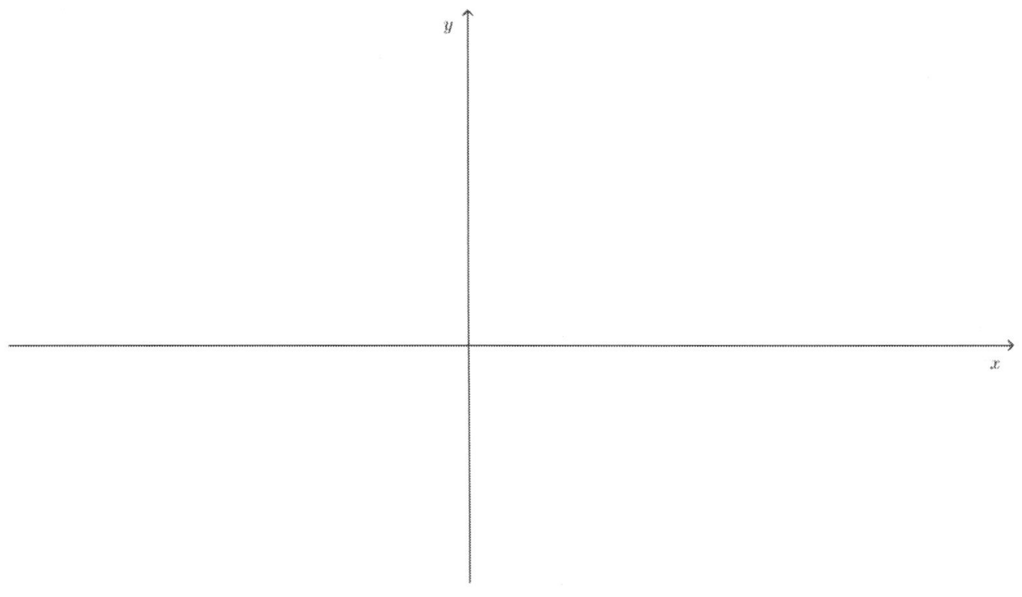

(b) What single geometric transformation best describes the change from the first graph to the second?

---

59. *(Non-calculator)* The numbers $1, 3$ and $x$ are written on a board. Their range is equal to the sum of their median and their mean. Find all possible values of $x$

60. In the pyramid below, triangular faces $ABC$, $ACD$ and $BCD$ are all right-angled at $C$, with $ABC$ horizontal and $D$ vertically above $C$

$AC = 2$

$BC = 3$

$CD = 1$

Calculate, to one decimal place, the angle between the faces $ABC$ and $ABD$

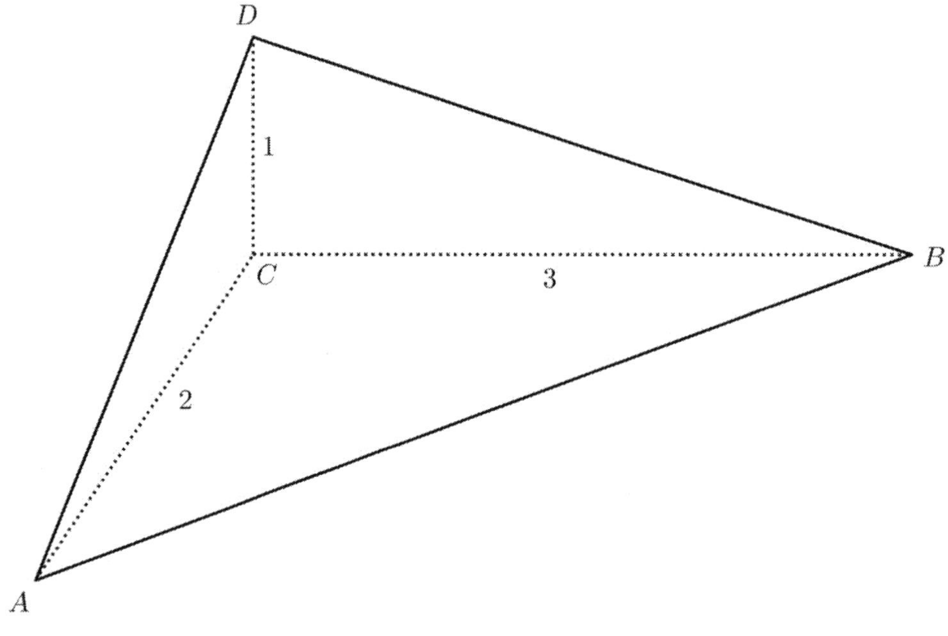

61. (a) A cube of side length 2 can be tightly packed with $1 \times 1 \times 2$ wooden blocks, all facing the same direction.

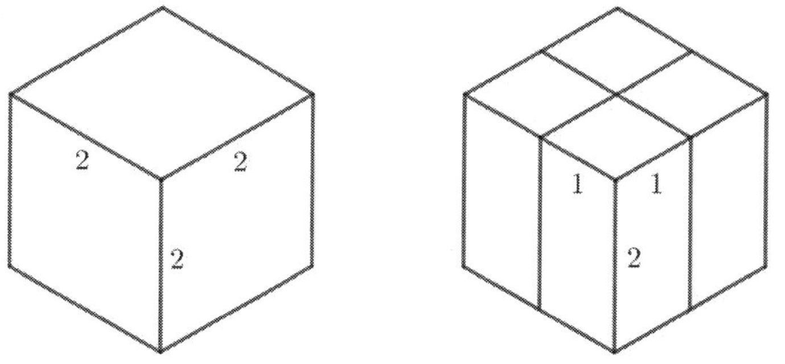


What is the side length of the smallest cube in which wooden blocks of dimensions $24 \times 42 \times 99$ can be tightly packed, all facing the same direction?

(b) A rectangle with dimensions $2 \times 4$ has the following properties:

- Its sides are in the same proportion as a $1 \times 2$ rectangle.
- It can be tightly packed with vertical $1 \times 2$ rectangles.
- It can be tightly packed with horizontal $1 \times 2$ rectangles.

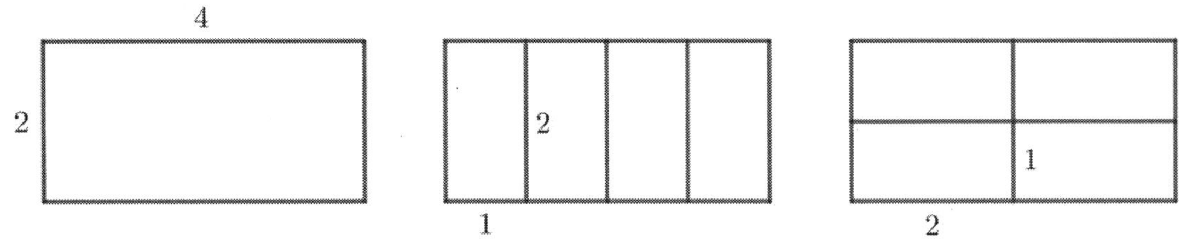


Find the dimensions of the smallest rectangle with all the following properties:

- Its sides are in the same proportion as a $10 \times 16$ rectangle.
- It can be tightly packed with vertical $10 \times 16$ rectangles.
- It can be tightly packed with horizontal $10 \times 16$ rectangles.

62. A collection of 50 identical blue marbles and 49 identical yellow marbles are to be split between two bags, with an odd number going into the green bag and an even number going into the red bag.

In how many ways can this be done?

63. Solve

$$\frac{x}{x+1} > 2$$

---

64. *(Non-calculator)* Solve

$$3^{4^n} = 9^{8^7}$$

(Note that $a^{b^c} = a^{(b^c)}$)

65. *(Non-calculator)* Solve

$$\sqrt[3]{\frac{4x}{15}} + 2 = \sqrt[3]{\frac{9x}{10}}$$

simplifying your answer as much as possible.

66. Below is a semicircle, centre $O$, of radius 9

$A, B$ and $C$ are points on the circumference such that $AC$ bisects $OB$ at point $P$, and angle $APB$ is twice angle $PAB$

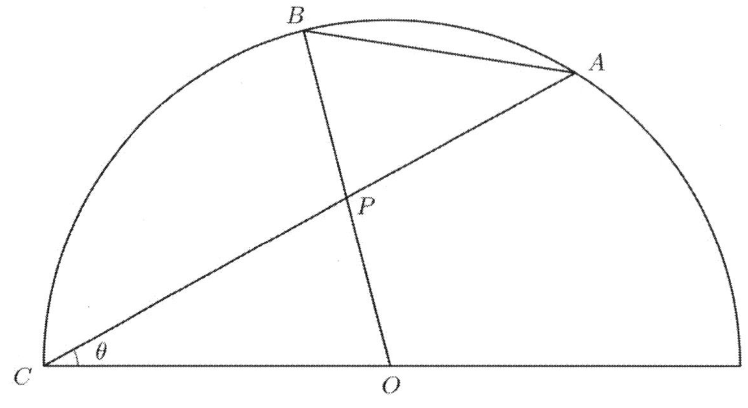


Calculate, to 2 decimal places:

(a) the size of angle $\theta$

(b) *(IGCSE)* the length $AP$

67.

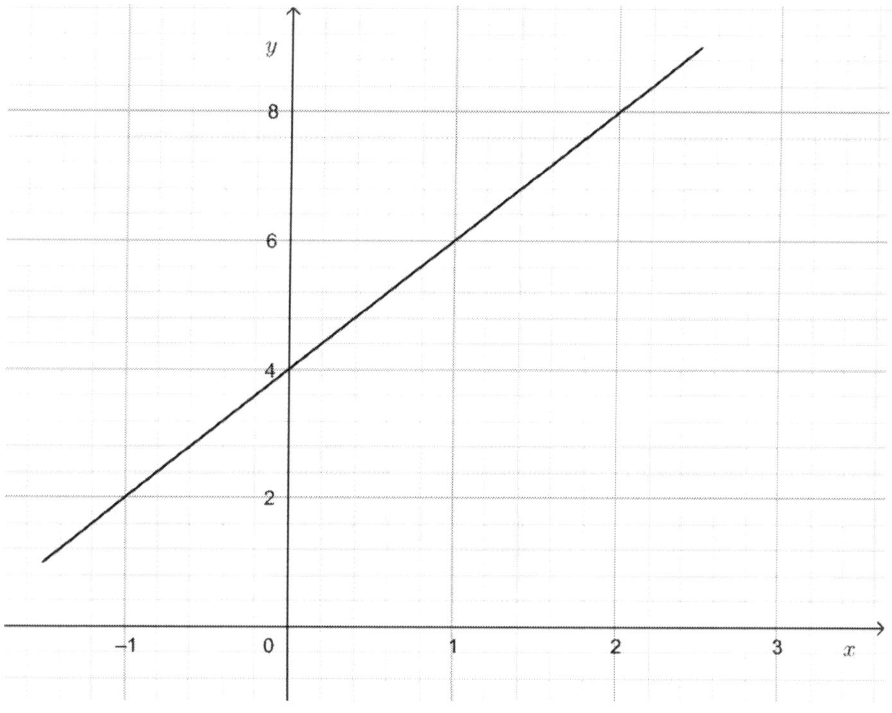


A point $P$ is chosen at random somewhere on the line segment above, between its endpoints. The coordinates of $P$ are $(p, q)$

Prove that the vectors $\begin{pmatrix} p+2 \\ q \end{pmatrix}$ and $\begin{pmatrix} p-3 \\ q-10 \end{pmatrix}$ point in opposite directions.

68. A student is trying to find all solutions to the equation

$$\sin(2x) = \frac{\sqrt{3}}{2}$$

in the range $0° \leq x \leq 360°$

He argues as follows:

$\sin^{-1}\left(\frac{\sqrt{3}}{2}\right) = 60°$, so one possibility is $2x = 60°$, or $x = 30°$

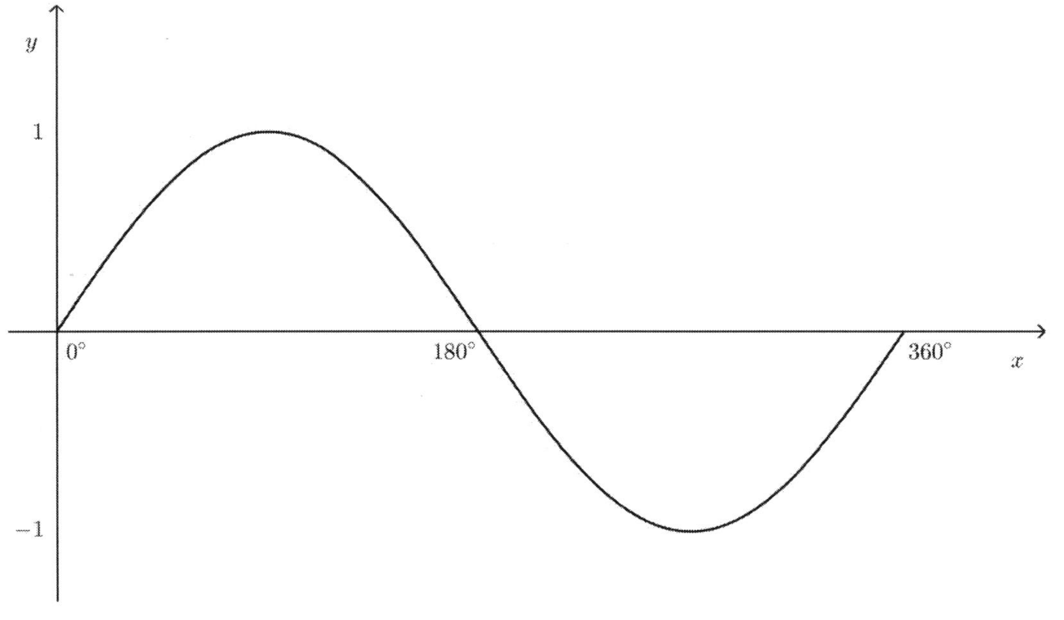


Using the sine graph, the other solution in the required range must be

$$x = 180° - 30° = 150°$$

So $x = 30°$ or $150°$

Find the flaw in the student's argument, and write a correct solution.

69. A triangle $T_0$ with sides of length 3, 4 and 5 is drawn.

Then six reflections are performed:

1. $T_0$ is reflected in its side of length 3 to form triangle $T_1$
2. $T_1$ is reflected in its side of length 4 to form triangle $T_2$
3. $T_2$ is reflected in its side of length 5 to form triangle $T_3$
4. $T_3$ is reflected in its side of length 3 to form triangle $T_4$
5. $T_4$ is reflected in its side of length 4 to form triangle $T_5$
6. $T_5$ is reflected in its side of length 5 to form triangle $T_6$

The first reflection has been shown on the next page.

(a) Prove that triangle $T_0$ can be transformed into triangle $T_6$ with a **single** transformation, and name this **type** of transformation.

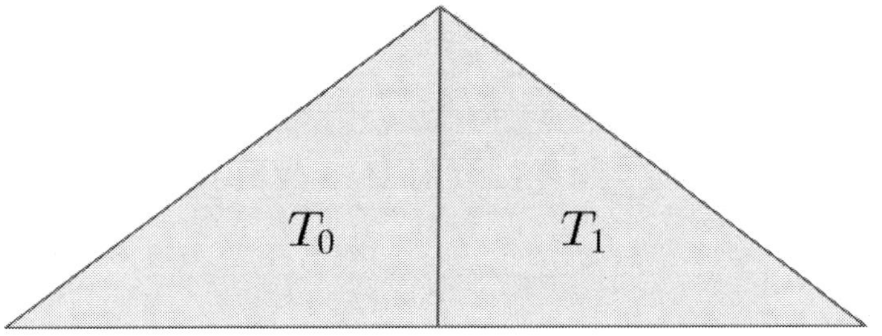
$T_0$
$T_1$

(b) A thin wooden triangle is positioned so that it covers $T_0$ precisely. It is then slid in one smooth motion across the page until it covers $T_6$ precisely, without the direction of motion ever changing.

Calculate the distance through which the wooden triangle has moved.

70. A thin beaker has a circular base of diameter 6cm, a circular rim of diameter 8cm and a height of 10cm. The centre of the rim is directly above the centre of the base.

Anna pours 250ml of orange juice into the beaker. Calculate the height of the surface of the liquid above the base of the beaker, giving your answer to 3 significant figures.

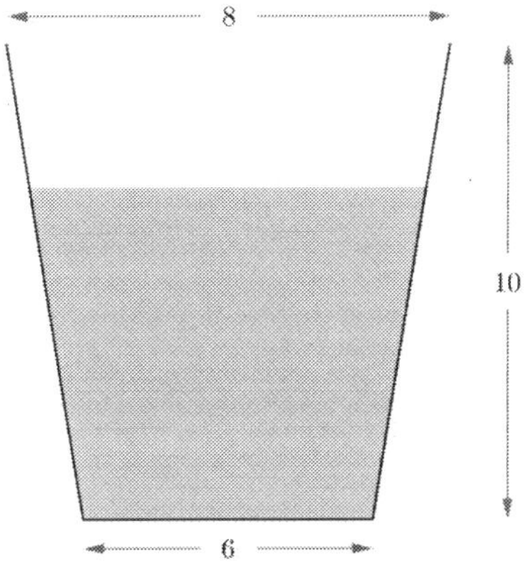

71. In Germainia every worker has a salary which is a whole number of pounds. Income tax is as follows:

- The first £10,000 is untaxed
- The next £40,000 is taxed at 20%
- Any amount above this is taxed at 50%

For example, someone earning a pre-tax salary of £120,000 would have to pay

0% of £10,000 + 20% of £40,000 + 50% of £70,000 = £43,000

in tax, leaving them with a post-tax salary of £77,000

In Noetherland, income is taxed differently:

- The first £15,000 is untaxed
- Any amount above this is taxed at 30%

Emmy lives in Germainia and Sophie lives in Noetherland. Their pre-tax salaries are equal, and their post-tax salaries are also equal.

What could their pre-tax salaries be? Find all possible values.

72. (a) In the quadrilateral below, the two marked sides are equal and the two marked angles are equal.

Prove that

(i) the two diagonals are equal in length,

(ii) the two unmarked sides are parallel.

(b) The diagram below shows a regular pentagon and part of a regular icosagon (with twenty sides).

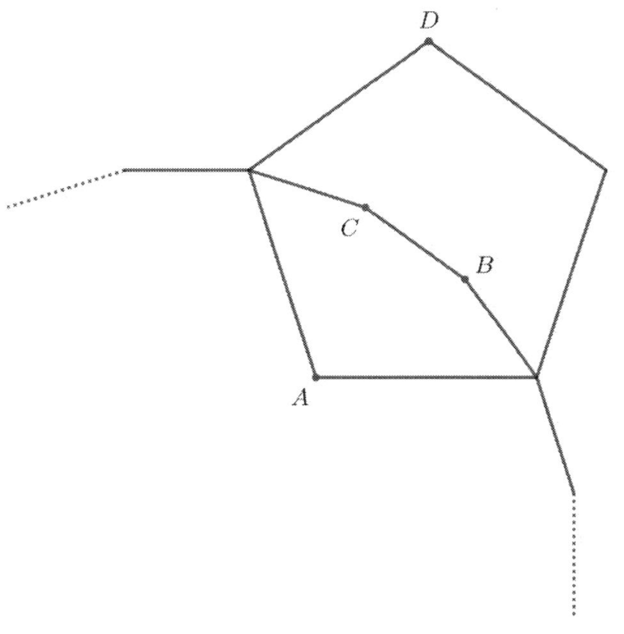


Prove that the distances $AB$ and $CD$ are equal.

73. Carlotta's son is helping her prepare a family meal. Carlotta can peel a potato in $p$ seconds and can chop a carrot in $q$ seconds. Her son won't be peeling any potatoes, but can chop a carrot in $r$ seconds.

In total, 10 potatoes need peeling and 24 carrots need chopping. Assuming they work together, find an expression in terms of $p, q$ and $r$ for the number of seconds it will take them to prepare all the potatoes and all the carrots, leaving your answer as a single simplified algebraic fraction.

(You may assume that Carlotta finishes peeling potatoes before all the carrots have been chopped, each carrot is chopped by only one person, and they finish chopping their final carrots at the same time.)

74.

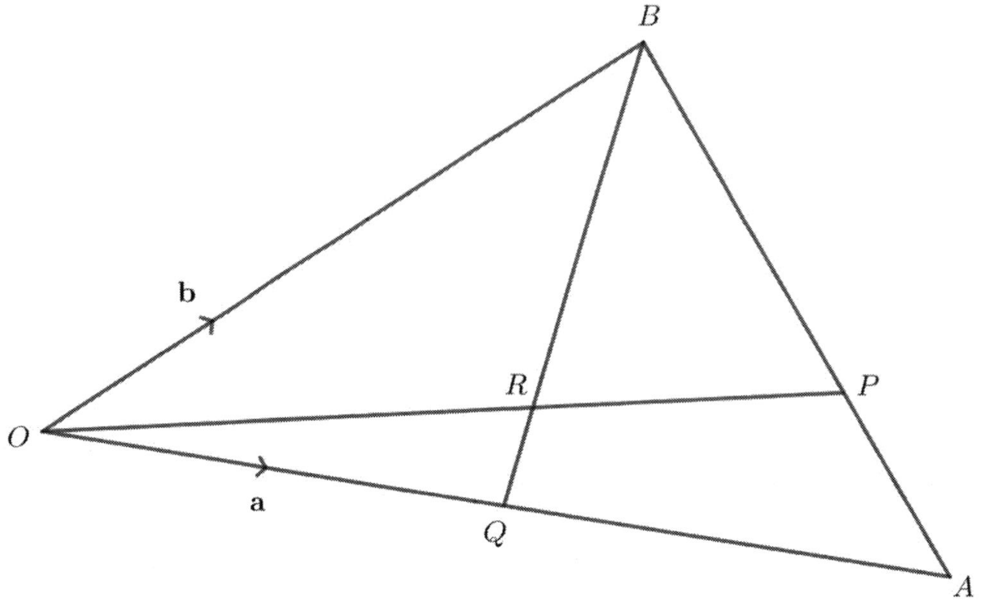


In triangle $OAB$ above, $\overrightarrow{OA} = \boldsymbol{a}$ and $\overrightarrow{OB} = \boldsymbol{b}$

$P$ is the point on $AB$ so that $AP : PB = 1 : 2$

$Q$ is the midpoint of $OA$

$R$ is the point of intersection of $OP$ and $BQ$

Express the following vectors in terms of $\boldsymbol{a}$ and $\boldsymbol{b}$:

(i) $\overrightarrow{AB}$

(ii) $\overrightarrow{OP}$

(iii) $\overrightarrow{BQ}$

(iv) $\overrightarrow{OR}$

75. A square pond of side length 2 metres is to be surrounded by twelve square paving stones of side length 1 metre.

The first design is shown below, constructed with a circle whose centre coincides with the centre of the pond.

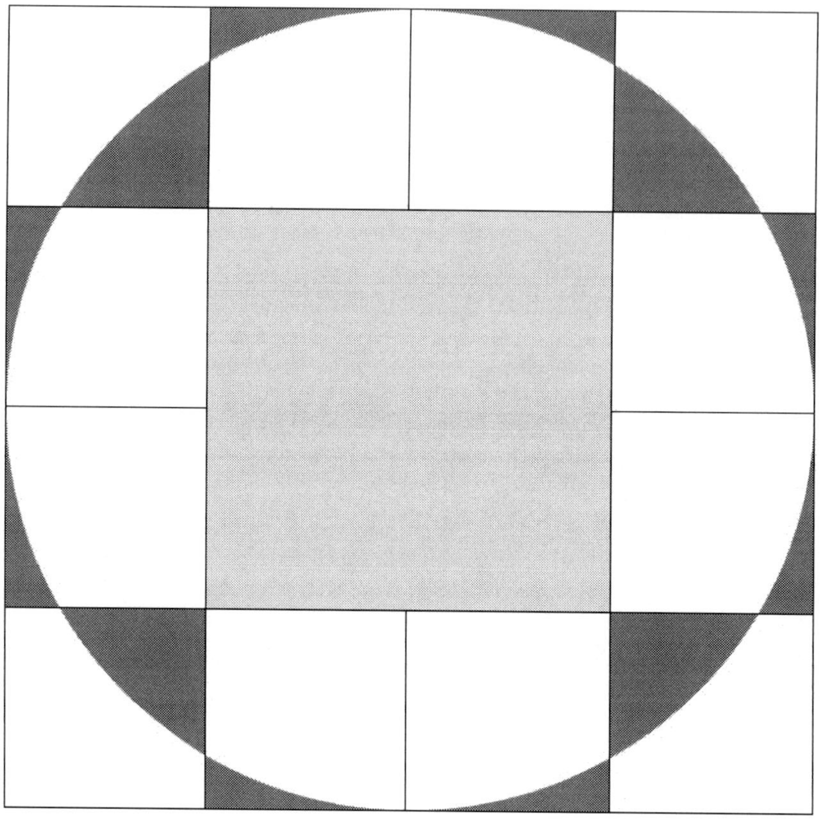

(a) Calculate exactly the total dark grey paved area for this design.

Here is a second design.

(b) Calculate exactly the total dark grey paved area for the second design.

# SOLUTIONS

1.

Mass upper bound is 10.25

Volume lower bound is 2.535

So the upper bound for the density is $\frac{10.25}{2.535} = \mathbf{4.04\ g/cm^3}$

Mass lower bound is 10.15

Volume upper bound is 2.545

So the lower bound for the density is $\frac{10.15}{2.545} = \mathbf{3.99\ g/cm^3}$

---

2.

(a)

$$(1.7 \times 10^{200}) - (1.8 \times 10^{199})$$
$$= (1.7 \times 10^{200}) - (0.18 \times 10^{200})$$
$$= (1.7 - 0.18) \times 10^{200}$$
$$= \mathbf{1.52 \times 10^{200}}$$

(b)

$$\left(4.3\times10^{-55}\right)\div\left(5\times10^{-122}\right)$$

$$=\frac{4.3\times10^{-55}}{5\times10^{-122}}$$

$$=\frac{4.3}{5}\times\frac{10^{-55}}{10^{-122}}$$

$$=0.86\times10^{67}$$

$$=\mathbf{8.6\times10^{66}}$$

---

3.

$$\sqrt[3]{\frac{11.8\times30.1}{2.94}}$$

$$\approx\sqrt[3]{\frac{12\times30}{3}}$$

$$=\sqrt[3]{12\times10}$$

$$=\sqrt[3]{120}$$

$$\approx\sqrt[3]{125}$$

$$=\mathbf{5}$$

4.

$$p\left(3-\sqrt{q}\right)=3+\sqrt{q}$$

$$3p-p\sqrt{q}=3+\sqrt{q}$$

$$\sqrt{q}+p\sqrt{q}=3p-3$$

$$\sqrt{q}(1+p)=3p-3$$

$$\sqrt{q}=\frac{3p-3}{1+p}$$

$$\boldsymbol{q=\left(\frac{3p-3}{1+p}\right)^{2}}$$

---

5.

Let $x=0.47\dot{6}$

Then

$$100x=47.\dot{6}$$

$$1000x=476.\dot{6}$$

$$900x=429$$

$$x=\frac{429}{900}=\frac{143}{300}$$

Therefore

$$0.47\dot{6}=\mathbf{\frac{143}{300}}$$

6.

We can think of Isabella having 4 choices for the scarf (scarf 1, scarf 2, scarf 3 or no scarf). Similarly we can think of her having 3 choices for the hat.

So she has $3 \times 4 \times 4 \times 4 \times 3 = 3^2 \times 4^3 = 9 \times 64 = \mathbf{576}$ ways to select an outfit.

---

7.

(a) Mayu is incorrect. For example:

$$\frac{2}{3} \times \frac{3}{2} = 1$$

but neither $\frac{2}{3}$ nor $\frac{3}{2}$ is an integer.

So the answer is **no**.

(b) Naoki is also incorrect. For example, the lines $x = 3$ and $y = 4$ are perpendicular, but the product of their gradients is not -1 (the first line has infinite gradient, and the second line has a gradient of zero).

So again the answer is **no**.

8.

(a)

| Number of cups sold $(x)$ | Frequency |
|---|---|
| $0 \le x < 100$ | 7 |
| $100 \le x < 150$ | 15 |
| $150 \le x < 200$ | 22 |
| $200 \le x < 250$ | 27 |
| $250 \le x < 300$ | 19 |
| $300 \le x < 500$ | 10 |

The second class $(100 \le x < 150)$ has a frequency of 15, and covers an area of $5 \times 30 = 150$ small squares on the histogram.

As area is proportional to frequency, it follows that each café corresponds to 10 small squares.

The final class $(300 \le x < 500)$ covers an area of $20 \times 5 = 100$ small squares, so its frequency is 10

Using the fact that the total number of cafés is 100, we can now work out the frequency of the first class $(0 \le x < 100)$:

$$100 - (15 + 22 + 27 + 19 + 10) = 7$$

(b)

To complete the histogram, note that the first class $(0 \le x < 100)$ has a frequency of 7, so needs to cover an area of 70 small squares; 10 squares wide and 7 squares tall.

Similarly the fifth class $(250 \le x < 300)$ has a frequency of 19, so needs to cover an area of 190 small squares; 5 squares wide and 38 squares tall.

Frequency density

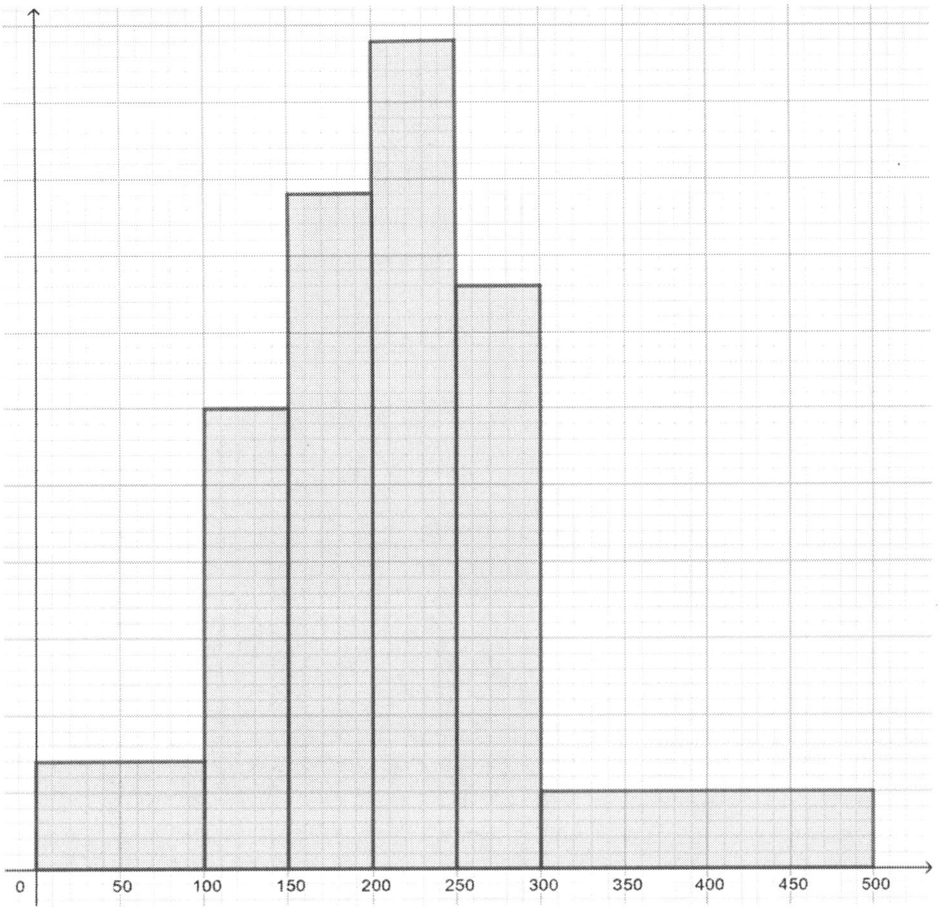


Number of coffee cups sold

---

9.

$$(y-2)(2y-3)(y^2+4)$$

$$= \left(2y^2 - 3y - 4y + 6\right)(y^2+4)$$

$$= \left(2y^2 - 7y + 6\right)(y^2+4)$$

$$= 2y^4 + 8y^2 - 7y^3 - 28y + 6y^2 + 24$$

$$= \mathbf{2y^4 - 7y^3 + 14y^2 - 28y + 24}$$

10.

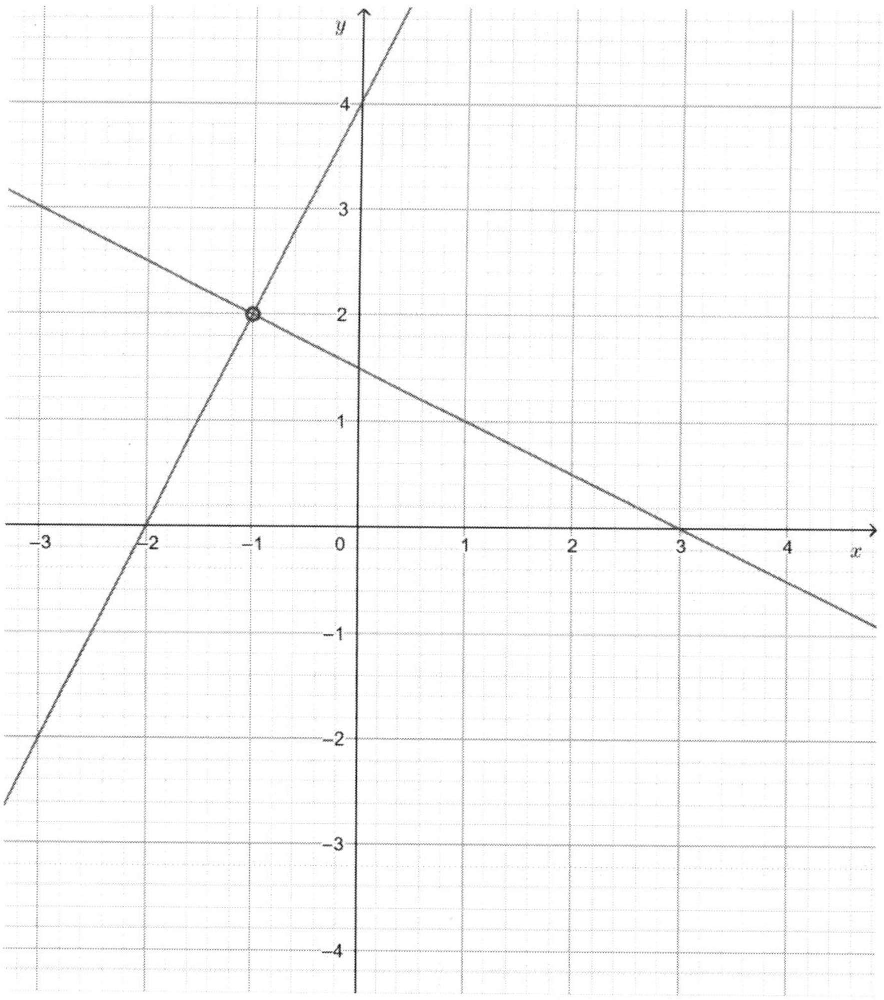


$$\boldsymbol{x = -1, \qquad y = 2}$$

---

11.

$$75ab^2 - 12a^3$$

$$= 3a(25b^2 - 4a^2)$$

$$= \mathbf{3}\boldsymbol{a}\mathbf{(5}\boldsymbol{b}\mathbf{ + 2}\boldsymbol{a}\mathbf{)(5}\boldsymbol{b}\mathbf{ - 2}\boldsymbol{a}\mathbf{)}$$

12. The integer $N$ must be a factor of 90

The factors of 90 are 1, 2, 3, 5, 6, 9, 10, 15, 18, 30, 45 and 90

Since $90 = 45 \times 2$, the lowest common multiple of $N$ and 45 will be 90 if $N$ is even, or 45 if $N$ is odd.

So the possible values of $N$ are $\mathbf{2, 6, 10, 18, 30}$ **and 90**

---

13. The number of dollars he receives before the trip is

$$250 \times 1.27 = 317.5$$

So the number of pounds he receives after the trip is

$$317.5 \div 1.37 = 231.7518\ldots$$

$$= 231.75 \text{ (nearest £0.01)}$$

Hence he has lost $250 - 231.75 = \text{£}\mathbf{18.25}$ as a result of the two exchanges.

---

14.

Let

$$F = \frac{k}{\sqrt[3]{OP}}$$

Then the information given in the question yields

$$15 = \frac{k}{\sqrt[3]{8}} = \frac{k}{2}$$

So $k = 30$. Hence, when $OP = F$, we have

$$OP = \frac{30}{\sqrt[3]{OP}}$$

$$OP^{\frac{4}{3}} = 30$$

$$OP = 30^{\frac{3}{4}} = \mathbf{12.8m}$$

15.

We need to compute

$$(9 + 18 + 27 \ldots + 900) - (7 + 14 + 21 + \cdots + 700)$$

Rearranging terms, this is the same as

$$(9 - 7) + (18 - 14) + (27 - 21) + \cdots + (900 - 700)$$

$$= 2 + 4 + 6 + \cdots + 200$$

This is an arithmetic series with first term 2, common difference 2 and 100 terms.

So the answer is

$$\frac{100}{2}(4 + 99 \times 2)$$

$$= 50 \times 202$$

$$= \mathbf{10100}$$

---

16. Work backwards in New York City local time:

| | |
|---|---|
| Arrive at the meeting | 8am Wed |
| Plane lands | 6am Wed |
| Plane takes off | 10pm Tues |
| Arrives at airport | 8pm Tues |
| Leaves home | 7:15pm Tues |

So she can leave her house at 7:15pm (NYC local time) on Tuesday 21st August.

Since London is 5 hours ahead of New York, the latest date and time she can leave her house is **12:15am** on **Wednesday 22nd August**.

17.

(i) $A \cup C'$

Numbers that are in $A$ or not in $C$:

**1, 2, 3, 4, 6, 7, 9, 10**

(ii) $(A' \cap B) \cup C$

Numbers that are in $B$ but not $A$, or in $C$:

**2, 3, 5, 6, 8**

---

18.

Let $x$ be the August price in pounds.

September price $= 0.6x$

October price $= 0.9 \times 0.6x = 0.54x$

So

$$x - 0.54x = 20.7$$

$$0.46x = 20.7$$

$$x = 45$$

Hence the September price $= 0.6 \times £45 =$ **£27**

19. With the given ratios, we must have

$$\left(\frac{n}{n+7}\right)\left(\frac{n}{n+10}\right)=\frac{n}{n+19}$$

$$\frac{n^2}{(n+7)(n+10)}=\frac{n}{n+19}$$

$$n^2(n+19)=n(n+7)(n+10)$$

$$n(n+19)=(n+7)(n+10)$$

$$n^2+19n=n^2+17n+70$$

$$2n=70$$

$$n=35$$

It follows that the ratio of cows to sheep is $35:42$, and the ratio of sheep to pigs is $35:45$, equivalent to $42:54$

Hence the ratio of cows to sheep to pigs is

$$35:42:54$$

and so the fraction of all animals which are sheep is

$$\frac{42}{35+42+54}=\mathbf{\frac{42}{131}}$$

---

20.

$$\frac{y}{xy+1}=\frac{1}{x+\frac{1}{y}}$$

$$=\frac{1}{3.125}$$

$$=\frac{1000}{3125}$$

$$=\frac{8}{25}$$

$$=\frac{32}{100}$$

$$=\mathbf{0.32}$$

21. Let the common ratio be $r$, so that the first five terms are $5, 5r, 5r^2, 5r^3$ and $5r^4$

The third term is 100, so

$$5r^2 = 100$$

$$r^2 = 20$$

$$r = \pm\sqrt{20} = \pm 2\sqrt{5}$$

But since the terms are all positive, we must have $r = 2\sqrt{5}$

Therefore the sum of the first five terms is

$$5 + 10\sqrt{5} + 100 + 200\sqrt{5} + 2000 = \mathbf{2105 + 210\sqrt{5}}$$

---

22.

$$\frac{38}{43 - x^4} = \frac{43}{38 - x^4}$$

Cross-multiply:

$$38(38 - x^4) = 43(43 - x^4)$$

$$38^2 - 38x^4 = 43^2 - 43x^4$$

$$5x^4 = 43^2 - 38^2$$

Use the difference of two squares:

$$5x^4 = (43 - 38)(43 + 38)$$

$$5x^4 = 5(81)$$

$$x^4 = 81$$

$$x = \pm\mathbf{3}$$

23.

(a) We have $5 + 10 + 15 + 20 = 50$, which is not divisible by 20

So the claim is **FALSE**.

(b) Let $4n$ be the smallest multiple of 4 in the sum.

We have

$$4n + (4n + 4) + (4n + 8) + (4n + 12) + (4n + 16)$$

$$= 20n + 40$$

$$= 20(n + 2)$$

Since $20(n + 2)$ is always divisible by 20, the claim is **TRUE**.

---

24. The volume scale factor is $\frac{45}{80} = 0.5625$

So the length scale factor is $\sqrt[3]{0.5625} = 0.82548 \ldots$

It follows that the area scale factor is $(0.82548 \ldots)^2 = 0.68142 \ldots$

Hence the smaller surface area is

$$240 \times 0.68142 \ldots$$

$$= \mathbf{164\ cm^2}$$

25.

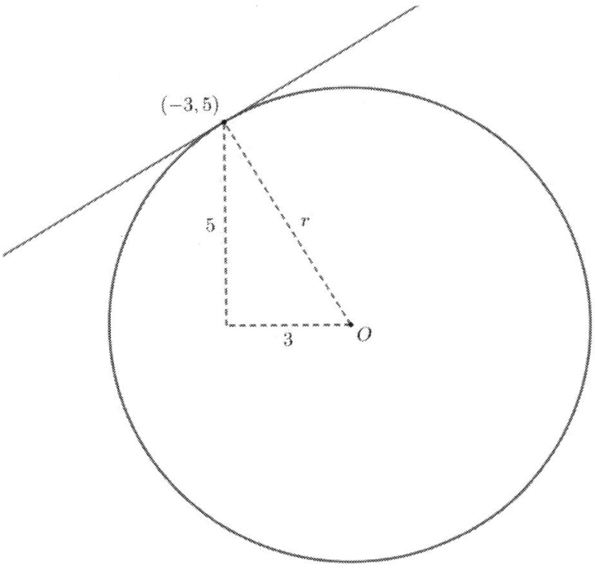


(a) Using Pythagoras, the radius $r$ must satisfy

$$r^2 = 3^2 + 5^2$$

$$r^2 = 9 + 25$$

$$r^2 = 34$$

Thus the equation of the circle is

$$\boldsymbol{x^2 + y^2 = 34}$$

(b) The tangent is perpendicular to the radius at the point of contact.

The gradient of the radius is

$$\frac{0-5}{3-0}=-\frac{5}{3}$$

So the gradient of the tangent is $\frac{3}{5}$

It follows that the equation of the tangent is

$$y=\frac{3}{5}x+c$$

for some value of $c$

The tangent passes through $(-3,5)$. Substituting this into the equation above yields

$$5=\frac{3}{5}(-3)+c$$

$$5=-\frac{9}{5}+c$$

$$c=5+\frac{9}{5}$$

$$=\frac{25}{5}+\frac{9}{5}$$

$$=\frac{34}{5}$$

Hence the tangent has equation

$$y=\frac{3}{5}x+\frac{34}{5}$$

26.

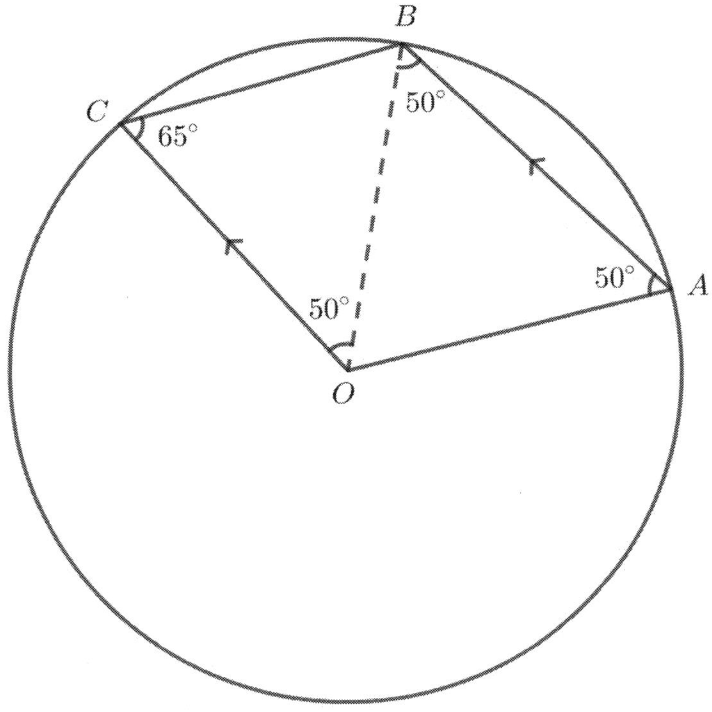


$OBA = 50°$ (triangle $OAB$ is isosceles)

$BOC = 50°$ ($OC$ and $AB$ are parallel)

$OCB = \frac{1}{2}(180° - 50°)$ (angles in a triangle sum to 180°, triangle $OCB$ is isosceles)

So $OCB =$ **65°**

27. Write the correct equation underneath each graph. Choose from the list of equations below.

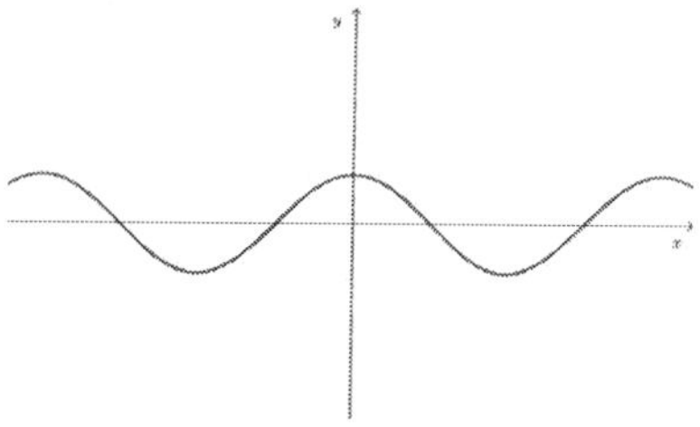

$y = \cos(x)$

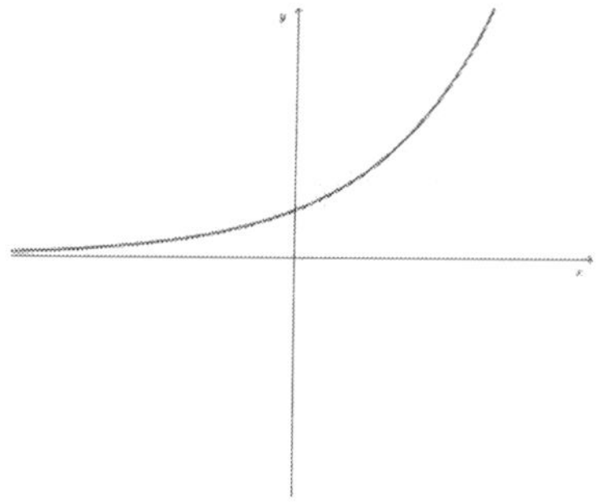

$y = \left(\frac{3}{2}\right)^x$

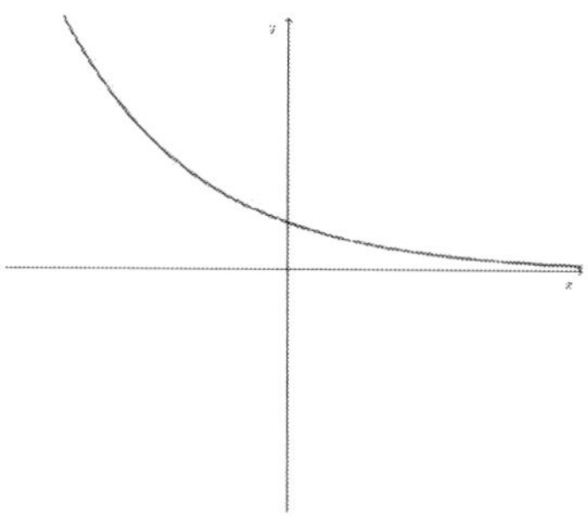

$y = 0.7^x$

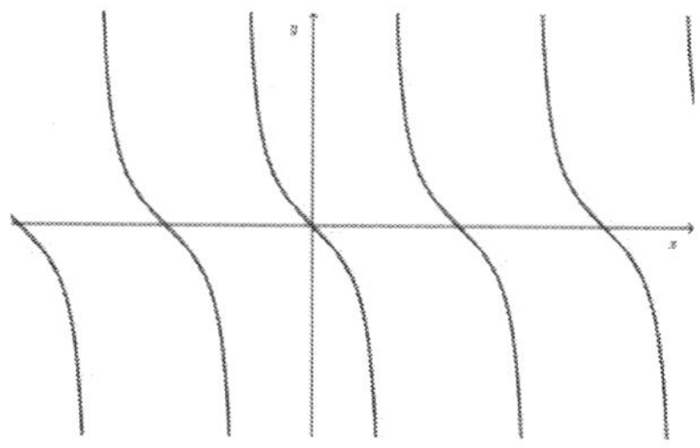

$y = -\tan(x)$

---

28. Mum takes $\frac{2}{5}$, leaving $\frac{3}{5}$

Dad takes $\frac{3}{5} \times \frac{3}{5} = \frac{9}{25}$, leaving $\frac{3}{5} - \frac{9}{25} = \frac{15}{25} - \frac{9}{25} = \frac{6}{25}$

So Matej receives $\frac{7}{15}$ of $\frac{6}{25}$ of the pizza, that is

$$\frac{7}{15} \times \frac{6}{25}$$

$$= \frac{7}{5} \times \frac{2}{25}$$

$$= \frac{14}{125}$$

$$= \frac{112}{1000}$$

$$= \mathbf{11.2\%}$$

29. Let $x$ be Lucas's original number. We have

$$\frac{15}{x} = 6x + 1$$

$$15 = x(6x + 1)$$

$$15 = 6x^2 + x$$

$$6x^2 + x - 15 = 0$$

$$(3x + 5)(2x - 3) = 0$$

$$x = -\frac{5}{3} \text{ or } \frac{3}{2}$$

But his number was positive; thus his original number was $\mathbf{\frac{3}{2}}$

---

30. Let the radius be $r$ cm

Then the quarter circle has area $\frac{1}{4}\pi r^2$ and the triangle has area $\frac{1}{2}r^2$

Using the area of the segment, we can form the equation

$$\frac{1}{2}r^2 + 1 = \frac{1}{4}\pi r^2$$

So

$$2r^2 + 4 = \pi r^2$$

$$\pi r^2 - 2r^2 = 4$$

$$(\pi - 2)r^2 = 4$$

$$r^2 = \frac{4}{\pi - 2}$$

Hence the area of the triangle is

$$\frac{1}{2}r^2 = \mathbf{\frac{2}{\pi-2}\ cm^2}$$

31.

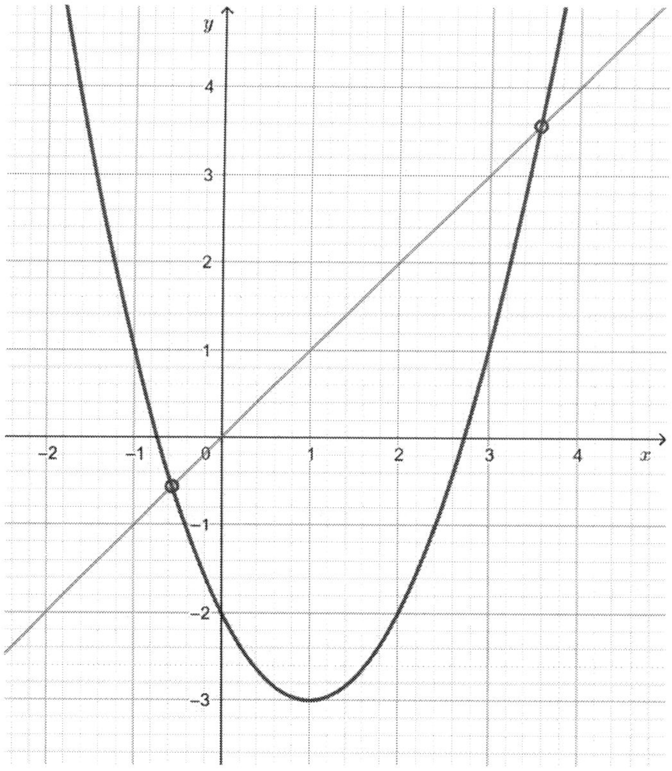


$$x + \frac{2}{x} = 3 + \frac{4}{x}$$

$$x^2 + 2 = 3x + 4$$

$$x^2 - 2x - 2 = x$$

So draw the line $y = x$, and find the $x$-coordinates of the points of intersection.

$\mathbf{x = -0.6}$ **or** $\mathbf{x = 3.6}$

32. Let $p$ be the probability of obtaining H on a single flip.

Then

$$p(1-p)+(1-p)p=0.42$$

$$2p(1-p)=0.42$$

$$2p-2p^2=0.42$$

$$2p^2-2p+0.42=0$$

$$p^2-p+0.21=0$$

Quadratic formula:

$$p=\frac{1\pm\sqrt{1-4\times1\times0.21}}{2}$$

$$p=\frac{1\pm\sqrt{0.16}}{2}$$

$$p=0.7 \text{ or } 0.3$$

But H is more likely than T, so $p=0.7$

Therefore $P(HH)=0.7^2=\mathbf{0.49}$

---

33.

The widths are $3,5,7,\ldots$ , increasing by 2 each time. So the width of the $n$th hammer is $2n+1$

The heights are $1,2,3,\ldots$ , increasing by 1 each time. So the height of the $n$th hammer is $n$

Finally, the handle lengths are also $1,2,3,\ldots$, increasing by 1 each time. So the length of the handle in the $n$th hammer is also $n$

Hence the number of dots in the $n$th hammer is

$$n(2n+1)+n$$

$$=n(2n+1+1)$$

$$=n(2n+2)$$

$$=\mathbf{2n(n+1)}$$

34.

(a)

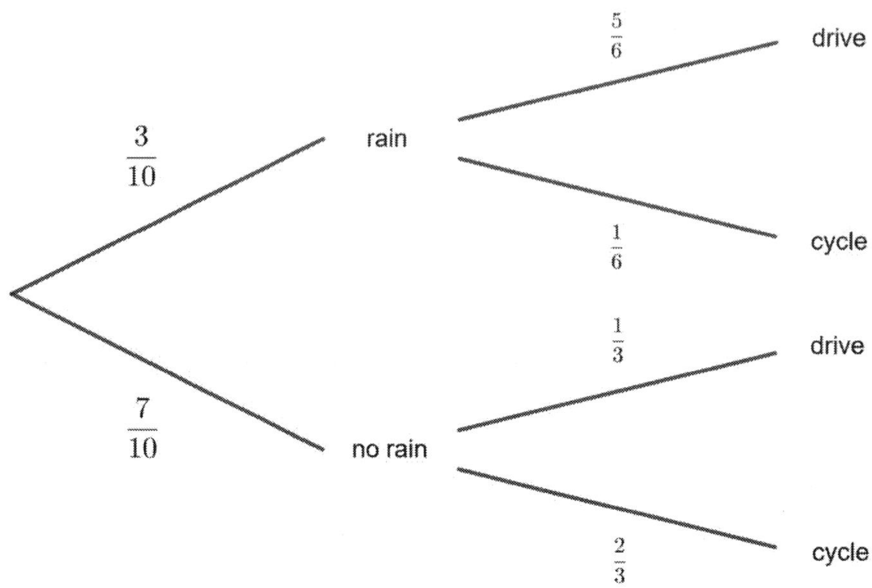


(b) On any given day, the probability that it rains and he drives is

$$\frac{3}{10}\times\frac{5}{6}=\frac{1}{4}$$

The probability that it does not rain and he drives is

$$\frac{7}{10}\times\frac{1}{3}=\frac{7}{30}$$

So of all the days on which he drives, we would expect the ratio of rainy days to dry days to be

$$\frac{1}{4}:\frac{7}{30}$$
$$=15:14$$

These numbers sum to 29. The number of driving days is 145, which is $5\times 29$

Hence we would expect it to have rained on $5\times 15=$ **75** of the days on which he drove into work.

35.

$$\frac{x-9}{x^2-9}-\frac{x-13}{x^2-2x-15}$$

$$=\frac{x-9}{(x+3)(x-3)}-\frac{x-13}{(x+3)(x-5)}$$

$$=\frac{(x-9)(x-5)}{(x+3)(x-3)(x-5)}-\frac{(x-13)(x-3)}{(x+3)(x-3)(x-5)}$$

$$=\frac{(x-9)(x-5)-(x-13)(x-3)}{(x+3)(x-3)(x-5)}$$

$$=\frac{\left(x^2-14x+45\right)-\left(x^2-16x+39\right)}{(x+3)(x-3)(x-5)}$$

$$=\frac{2x+6}{(x+3)(x-3)(x-5)}$$

$$=\frac{2(x+3)}{(x+3)(x-3)(x-5)}$$

$$=\mathbf{\frac{2}{(x-3)(x-5)}}$$

---

36. If we find the equation of the line through (1,2) and perpendicular to $3x + 4y = 5$, then we can solve the problem by finding the distance between (1,2) and the point where the lines intersect.

The equation

$$3x+4y=5$$

can be rearranged to

$$y=-\frac{3}{4}x+\frac{5}{4}$$

So its gradient is $-\frac{3}{4}$, and the gradient of the perpendicular line will be $\frac{4}{3}$, with equation

$$y=\frac{4}{3}x+c$$

Since this line passes through $(1,2)$, we have

$$2 = \frac{4}{3} + c$$

$$c = \frac{2}{3}$$

Hence the perpendicular through (1,2) has equation

$$y = \frac{4}{3}x + \frac{2}{3}$$

Setting this equal to the original line yields

$$\frac{4}{3}x + \frac{2}{3} = -\frac{3}{4}x + \frac{5}{4}$$

$$16x + 8 = -9x + 15$$

$$25x = 7$$

$$x = \frac{7}{25}$$

Substituting into $y = \frac{4}{3}x + \frac{2}{3}$ gives

$$y = \frac{4}{3} \times \frac{7}{25} + \frac{2}{3} = \frac{26}{25}$$

It follows that the two lines intersect at $(\frac{7}{25}, \frac{26}{25})$, and now Pythagoras gives us the required distance:

$$\left(1 - \frac{7}{25}\right)^2 + \left(2 - \frac{26}{25}\right)^2 = 1.44$$

and so the answer is

$$\sqrt{1.44} = \mathbf{1.2}$$

37.

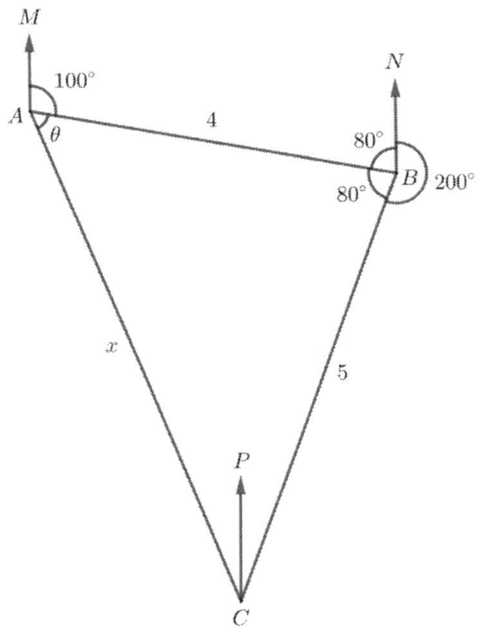


In the diagram above, $A$ is the starting point.

$AM$ and $BN$ are parallel, so $ABN = 180° - 100° = 80°$

Therefore $ABC = 360° - 80° - 200° = 80°$

So by the cosine rule,

$$x^2 = 4^2 + 5^2 - 2 \times 4 \times 5 \times \cos 80°$$

$$x^2 = 34.054 \ldots$$

$$x = 5.8355 \ldots$$

Now by the sine rule, $\frac{\sin\theta}{5} = \frac{\sin 80°}{x}$

So $\sin\theta = \frac{5\sin 80°}{5.8355\ldots} = 0.8438 \ldots$

Hence $\theta = \sin^{-1}(0.8438 \ldots) = 57.54 \ldots °$ or $\theta = 180° - \sin^{-1}(0.8438 \ldots) = 122.4 \ldots °$

But $ABC = 80°$, so $\theta = 57.54 \ldots °$

So $ACP = 180° - 100° - \theta = 22.45 \ldots °$

Therefore the bearing of $A$ from $C$ is $360° - 22.45 \ldots ° = 337.5 \ldots °$

Armani must walk for **5.84 km** on a bearing of **338°**

38. (a)

$$f(1.8)$$

$$= \frac{1}{1.8 - 2}$$

$$= \frac{1}{-0.2}$$

$$= \mathbf{-5}$$

(b)

$$g(2x - 1)$$

$$= (2x - 1)^2 + 4(2x - 1)$$

$$= 4x^2 - 2x - 2x + 1 + 8x - 4$$

$$= \mathbf{4x^2 + 4x - 3}$$

(c)

$$gf(x) = 0$$

$$g\left(\frac{1}{x - 2}\right) = 0$$

$$\left(\frac{1}{x - 2}\right)^2 + 4\left(\frac{1}{x - 2}\right) = 0$$

$$\frac{1}{(x - 2)^2} + \frac{4}{x - 2} = 0$$

Multiplying both sides by $(x - 2)^2$ gives

$$1 + 4(x - 2) = 0$$

$$1 + 4x - 8 = 0$$

$$4x = 7$$

$$\mathbf{x = \frac{7}{4}}$$

39. Let $x$ be the side length of the sheet of paper. We can write down the sides of the rectangle in terms of $x$ as below.

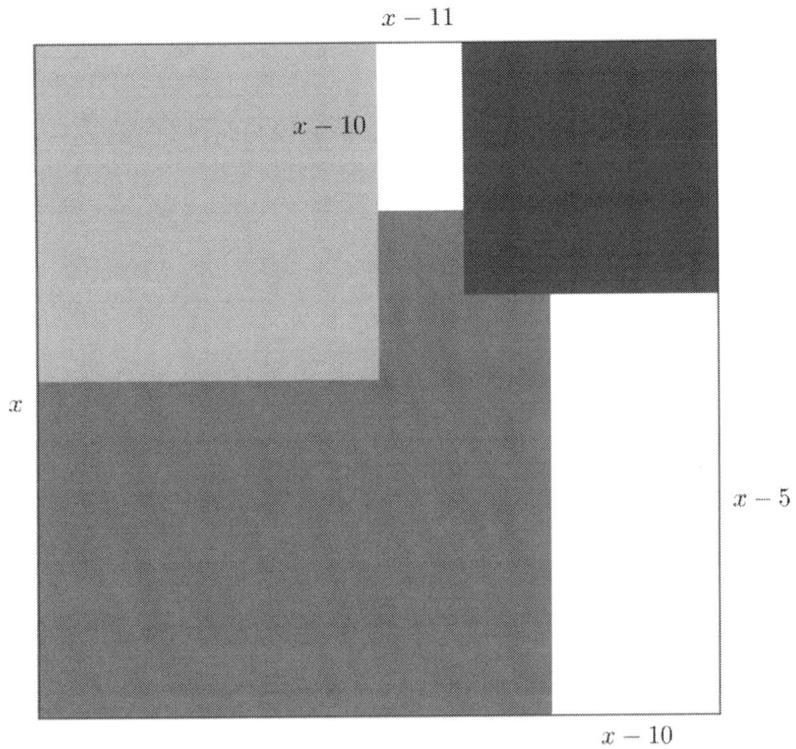


The information given means that

$$(x-5)(x-10)-(x-11)(x-10)=17$$

$$\big((x-5)-(x-11)\big)(x-10)=17$$

$$6(x-10)=17$$

$$x-10=\frac{17}{6}$$

$$x-10=2.8\dot{3}$$

$$x=12.8\dot{3}$$

Hence the side length of the square is $\mathbf{12.8\dot{3}}$ **cm**

40.

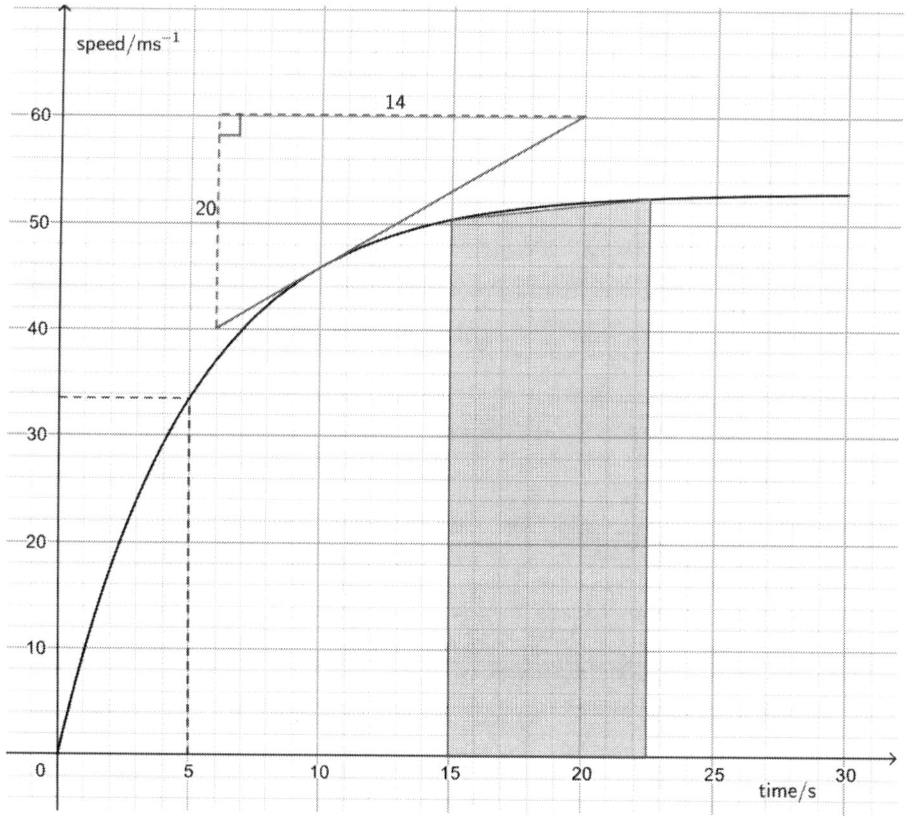


(a) **34 ms$^{-1}$**

(b) Draw a tangent to the curve at 10 seconds, and complete a right-angled triangle. The gradient of the tangent is approximately

$$\frac{20}{14} = \frac{10}{7} = 1.428\ldots$$

and so the skydiver's acceleration is approximately **1.4 ms$^{-2}$**

(c) We need to estimate the area beneath the curve. Split the area under the graph into four strips of equal width, drawing vertical lines at times of 7.5, 15, 22.5 and 30 seconds. One such strip is shown above.

This gives us one triangle and three trapezia, with a total area of approximately

$$\frac{1}{2}(7.5)(41) + \frac{1}{2}(7.5)(41 + 50) + \frac{1}{2}(7.5)(50 + 52) + \frac{1}{2}(7.5)(52 + 53)$$

$$= 1271.25$$

Hence to two significant figures, our estimate for the total distance travelled in the first 30 seconds is

**1300 metres**

41.

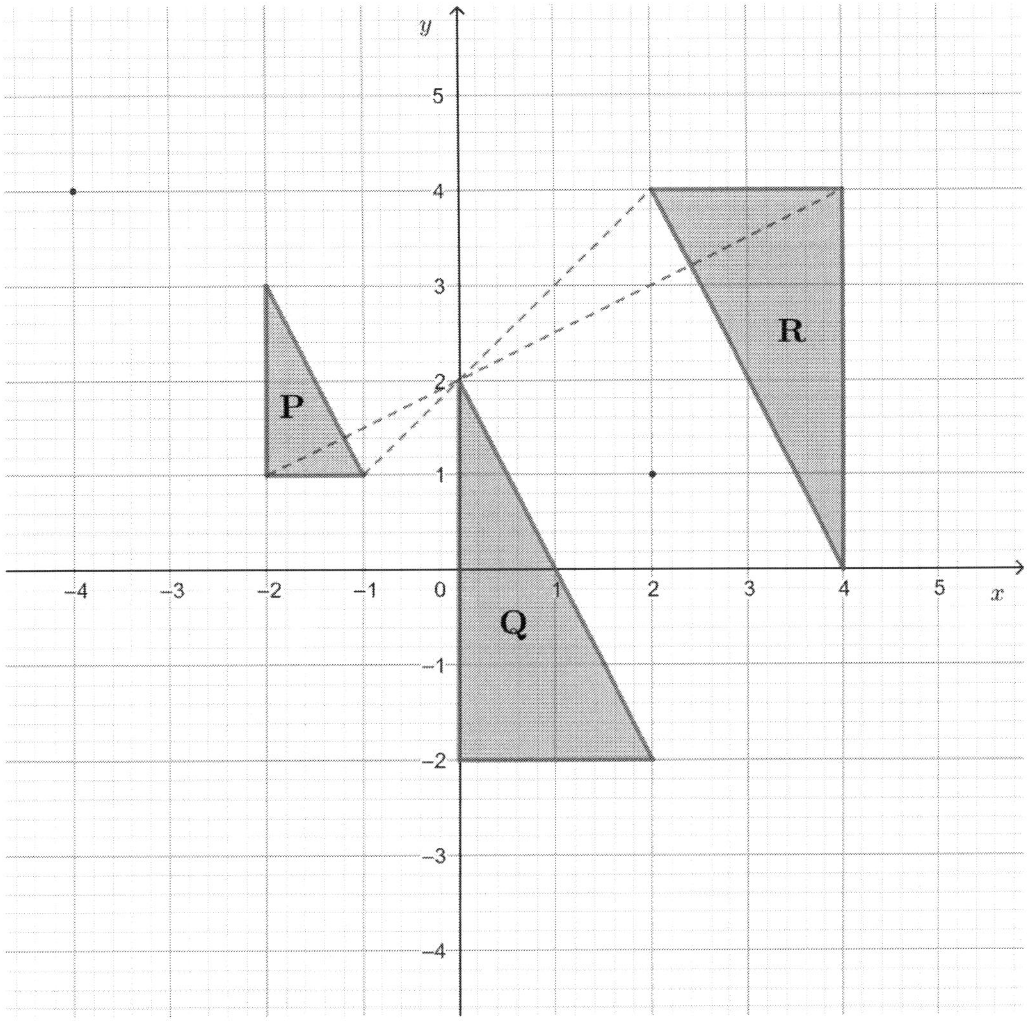


**Enlargement, scale factor $-\frac{1}{2}$, centre $(0,2)$**

42.

$$2x = -2 - 3y$$

Also,

$$2x(8 - 3y) = 56$$

So

$$(-2 - 3y)(8 - 3y) = 56$$

$$-16 + 6y - 24y + 9y^2 = 56$$

$$9y^2 - 18y - 72 = 0$$

$$y^2 - 2y - 8 = 0$$

$$(y + 2)(y - 4) = 0$$

So $y = -2$ or 4

Recall that $2x = -2 - 3y$

Hence when $y = -2$, we have

$$2x = -2 + 6 = 4$$

$$x = 2$$

When $y = 4$, we have

$$2x = -2 - 12 = -14$$

$$x = -7$$

Therefore $\mathbf{x = 2, y = -2}$ or $\mathbf{x = -7, y = 4}$

43. Draw a rectangle around the triangle, then subtract the unwanted areas from the total area of the rectangle.

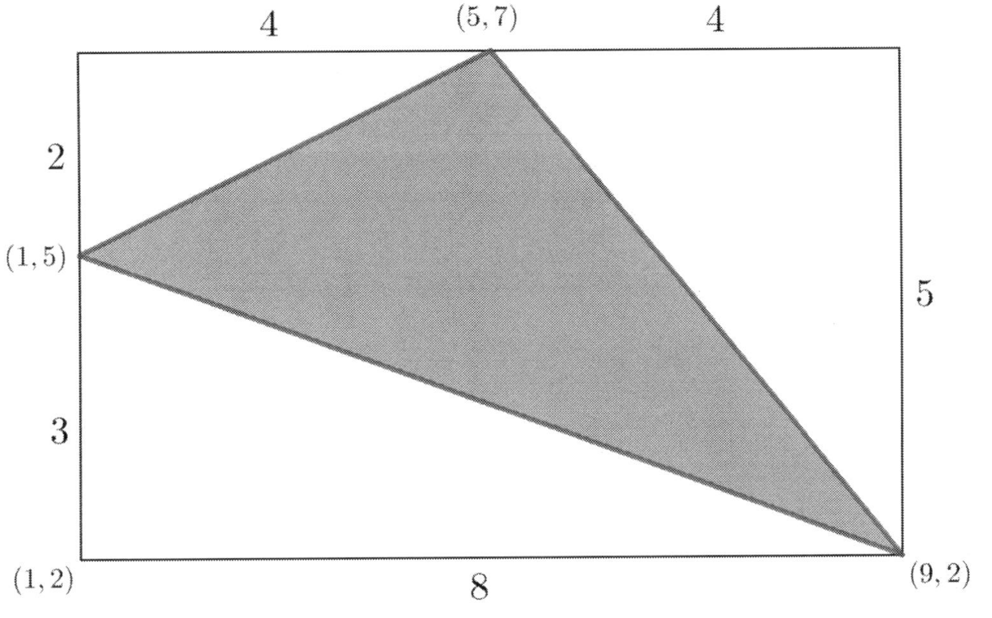


The required area is

$$(8 \times 5) - \left(\frac{1}{2} \times 8 \times 3\right) - \left(\frac{1}{2} \times 4 \times 5\right) - \left(\frac{1}{2} \times 4 \times 2\right)$$

$$= 40 - 12 - 10 - 4$$

$$= \mathbf{14}$$

44.

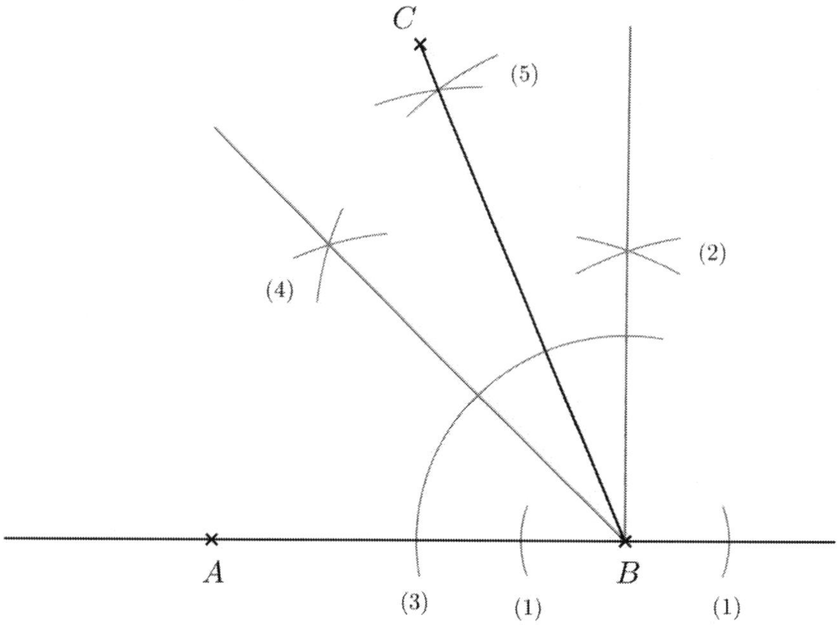


Construct a perpendicular to $AB$ at $B$

Then construct an angle bisector to cut the $90°$ angle into two $45°$ angles.

Then bisect one of the $45°$ angles to create two $22.5°$ angles.

Now we have

$$ABC = 45° + 22.5° = 67.5°$$

as required.

45.

When $t = 1$, $N = 2400$, and when $t = 3$, $N = 5400$

So $2400 = ar$ and $5400 = ar^3$

Dividing the second equation by the first yields

$$\frac{9}{4} = r^2$$

Hence $r = \frac{3}{2}$ (as $r$ is positive), and then the first equation gives us $a = 1600$

Therefore the number of cells at $t = 10$ is predicted to be

$$1600 \times \left(\frac{3}{2}\right)^{10} = 92264.06\ldots$$

$$= \mathbf{92300} \text{ (to the nearest hundred)}$$

---

46.

$$\sqrt[5]{2^n} = \sqrt{4^{n+1}}$$

$$(2^n)^{\frac{1}{5}} = \left(4^{n+1}\right)^{\frac{1}{2}}$$

$$2^{\frac{n}{5}} = 4^{\frac{n+1}{2}}$$

$$2^{\frac{n}{5}} = \left(2^2\right)^{\frac{n+1}{2}}$$

$$2^{\frac{n}{5}} = 2^{2\left(\frac{n+1}{2}\right)}$$

$$2^{\frac{n}{5}} = 2^{n+1}$$

$$\frac{n}{5} = n + 1$$

$$n = 5n + 5$$

$$-5 = 4n$$

$$\boldsymbol{n = -\frac{5}{4}}$$

47.

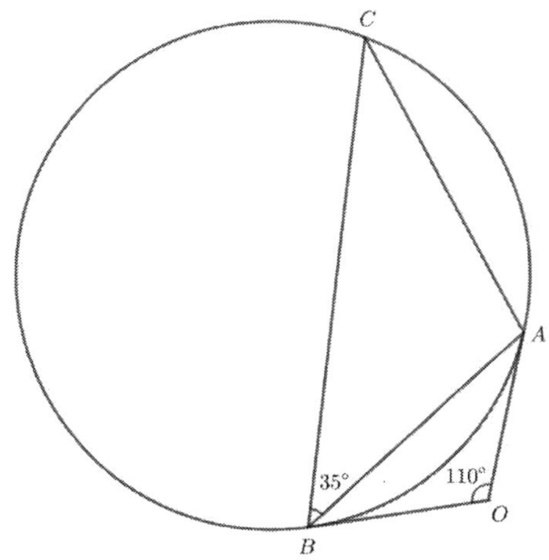


(a) We have $OA = OB$ (tangents from the same point)

So $OAB = OBA$ (base angles in an isosceles triangle)

$= 35°$ (angles in a triangle sum to $180°$ )

Hence $OAB = ABC$ are alternate angles, and thus $OA$ is parallel to $BC$

(b) By the alternate segment theorem,

$$ACB = OAB = 35°$$

So $BAC = 110°$, and triangles $OAB$ and $ABC$ are similar (they have the same angles).

Hence their sides are proportional, and so

$$\frac{OA}{AB} = \frac{AB}{BC}$$

Therefore

$$OA \times BC = AB^2$$

48.

$$\frac{\sqrt{8}}{4}+\frac{6}{\sqrt{32}}$$

$$=\frac{\sqrt{4}\sqrt{2}}{4}+\frac{6}{\sqrt{16}\sqrt{2}}$$

$$=\frac{2\sqrt{2}}{4}+\frac{6}{4\sqrt{2}}$$

$$=\frac{\sqrt{2}}{2}+\frac{3}{2\sqrt{2}}$$

$$=\frac{\sqrt{2}}{2}+\frac{3\sqrt{2}}{2\sqrt{2}\sqrt{2}}$$

$$=\frac{\sqrt{2}}{2}+\frac{3\sqrt{2}}{4}$$

$$=\frac{2\sqrt{2}}{4}+\frac{3\sqrt{2}}{4}$$

$$=\frac{5\sqrt{2}}{4}$$

$$=\boldsymbol{\frac{5}{4}\sqrt{2}}$$

---

49.

Either (i) $P$ lies between $(3,1)$ and $(9,10)$ , or (ii) $(9,10)$ lies between $(3,1)$ and $P$

(i) $P$ must lie two thirds of the way from $(3,1)$ to $(9,10)$. So $P = (7,7)$

(ii) $(9,10)$ must be the midpoint of $(3,1)$ and $P$. So $P = (15,19)$

Hence $\boldsymbol{P = (7,7)}$ **or** $\boldsymbol{(15,19)}$

50. Let the sphere have radius $r$, and let the cone have perpendicular height $h$ and slant height $l$

Then considering volumes we have

$$\frac{1}{3}\pi r^2 h = \frac{4}{3}\pi r^3$$

$$h = 4r$$

Now considering surface areas we have

$$4k\pi r^2 = \pi r^2 + \pi r l$$

$$4kr = r + l$$

Pythagoras gives us

$$4kr = r + \sqrt{r^2 + h^2}$$

But $h = 4r$ and so

$$4kr = r + \sqrt{r^2 + 16r^2}$$

$$4kr = r + \sqrt{17r^2}$$

$$4kr = r + r\sqrt{17}$$

$$k = \frac{\mathbf{1 + \sqrt{17}}}{\mathbf{4}}$$

51.

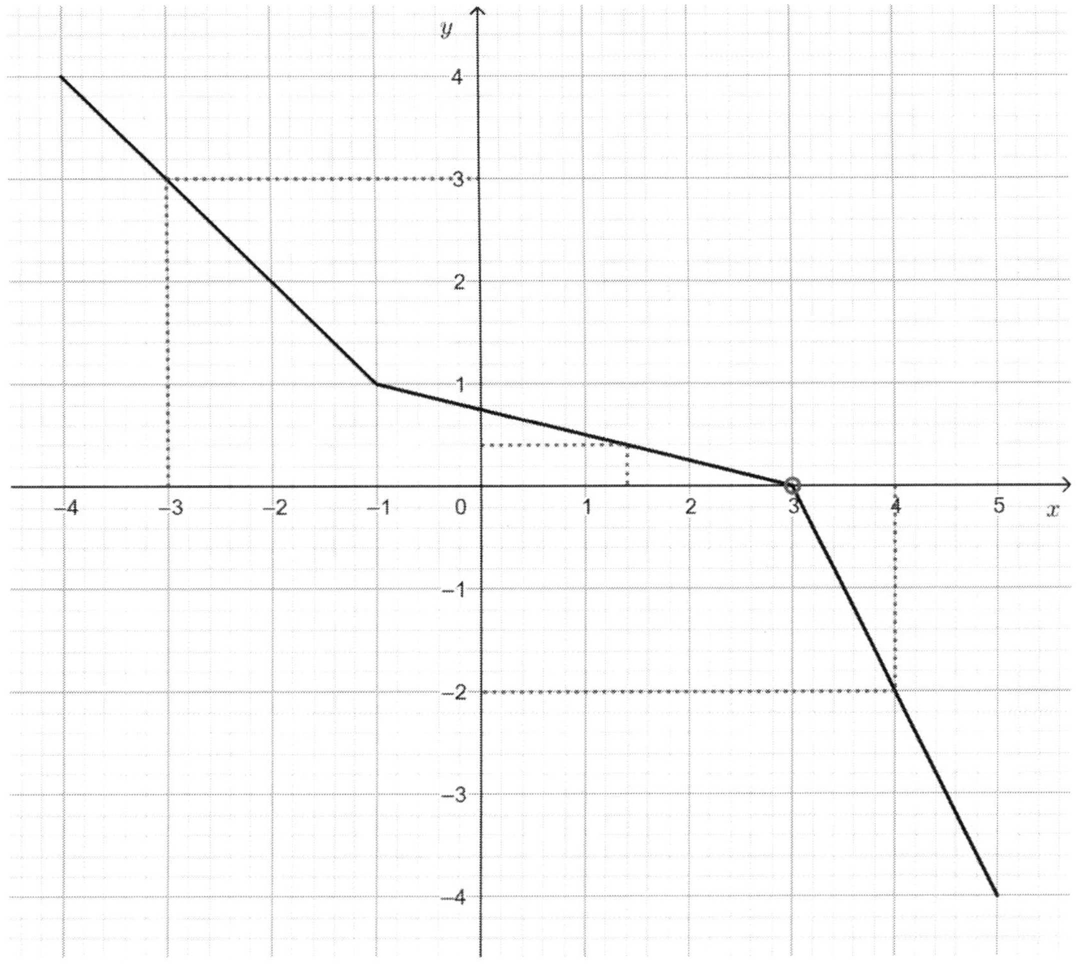


(a) The domain is $x \in \mathbb{R}, \mathbf{-4 \leq x \leq 5}$

(b) The range is $f(x) \in \mathbb{R}, \mathbf{-4 \leq f(x) \leq 4}$

(c) $f(4) = \mathbf{-2}$

(d) $ff(-3) = f\big(f(-3)\big) = f(3) = \mathbf{0}$

(e) $f^{-1}(0.4) = \mathbf{1.4}$

52. (a)

$$9^{\frac{1}{3}} \div 9^{\frac{5}{6}}$$

$$= 9^{\frac{2}{6}} \div 9^{\frac{5}{6}}$$

$$= 9^{\frac{2}{6} - \frac{5}{6}}$$

$$= 9^{-\frac{3}{6}}$$

$$= 9^{-\frac{1}{2}}$$

$$= \frac{1}{9^{\frac{1}{2}}}$$

$$= \mathbf{\frac{1}{3}}$$

(b)

$$\left(\sqrt[37]{5}\right)^{111}$$

$$= \left(5^{\frac{1}{37}}\right)^{111}$$

$$= 5^{\frac{111}{37}}$$

$$= 5^3$$

$$= \mathbf{125}$$

(c)

$$16^{1.75}$$

$$= 16^{\frac{7}{4}}$$

$$= \left(16^{\frac{1}{4}}\right)^7$$

$$= 2^7$$

$$= \mathbf{128}$$

53. (a)

$$f(x) = x^2 + 8x + 23$$

$$= (x+4)^2 - 16 + 23$$

$$= \mathbf{(x+4)^2 + 7}$$

(b)

$$\mathbf{(-4, 7)}$$

(c)

$$y = (x+4)^2 + 7$$

$$(x+4)^2 = y - 7$$

$$x + 4 = \sqrt{y-7}$$

$$x = \sqrt{y-7} - 4$$

Hence $\boldsymbol{f^{-1}(x) = \sqrt{x-7} - 4}$

(d)

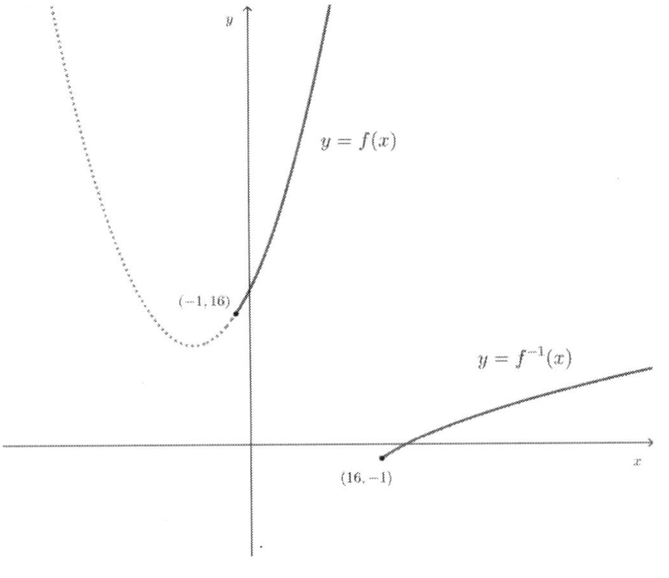


$$\mathbf{x \geq 16}$$

54.

$$(x+1)(x+2) < 12$$

$$x^2 + 2x + x + 2 < 12$$

$$x^2 + 3x - 10 < 0$$

$$(x+5)(x-2) < 0$$

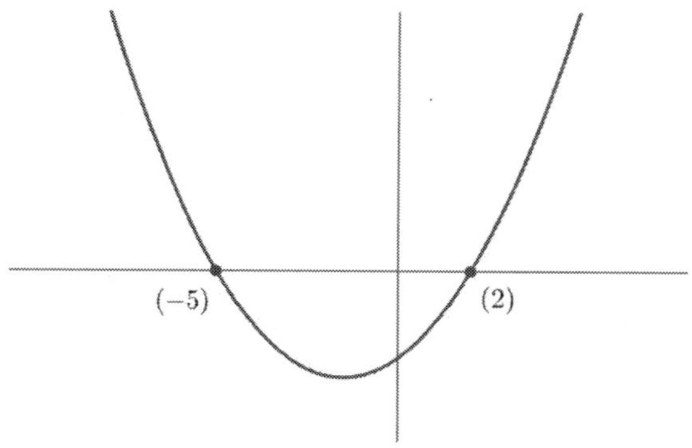


So the solution is $\mathbf{-5} < x < 2$

---

55.

(a)

$$\left(x+\sqrt[4]{x}\right)\left(x-\sqrt[4]{x}\right) = 2$$

$$x^2 - \sqrt{x} = 2$$

$$x^2 = 2 + \sqrt{x}$$

$$x = \sqrt{2+\sqrt{x}}$$

(b)

$$x_1 = \sqrt{2 + \sqrt{x_0}}$$

$$= \sqrt{2 + \sqrt{1}}$$

$$= \sqrt{3}$$

$$= \mathbf{1.7321}$$

$$x_2 = \sqrt{2 + \sqrt{x_1}}$$

$$= \mathbf{1.8210}$$

$$x_3 = \sqrt{2 + \sqrt{x_2}}$$

$$= \mathbf{1.8301}$$

(c)

$$\left(1.825 + \sqrt[4]{1.825}\right)\left(1.825 - \sqrt[4]{1.825}\right) = 1.97 \ldots < 2$$

$$\left(1.835 + \sqrt[4]{1.835}\right)\left(1.835 - \sqrt[4]{1.835}\right) = 2.01 \ldots > 2$$

So $x = \mathbf{1.83}$ is correct to three significant figures.

56. Rearrange the four inequalities so that they can be more easily represented on the axes:

$$x > -\frac{1}{2}$$

$$x < \frac{3}{4}$$

$$y < -x + 5$$

$$y > -\frac{6}{7}x + \frac{8}{7}$$

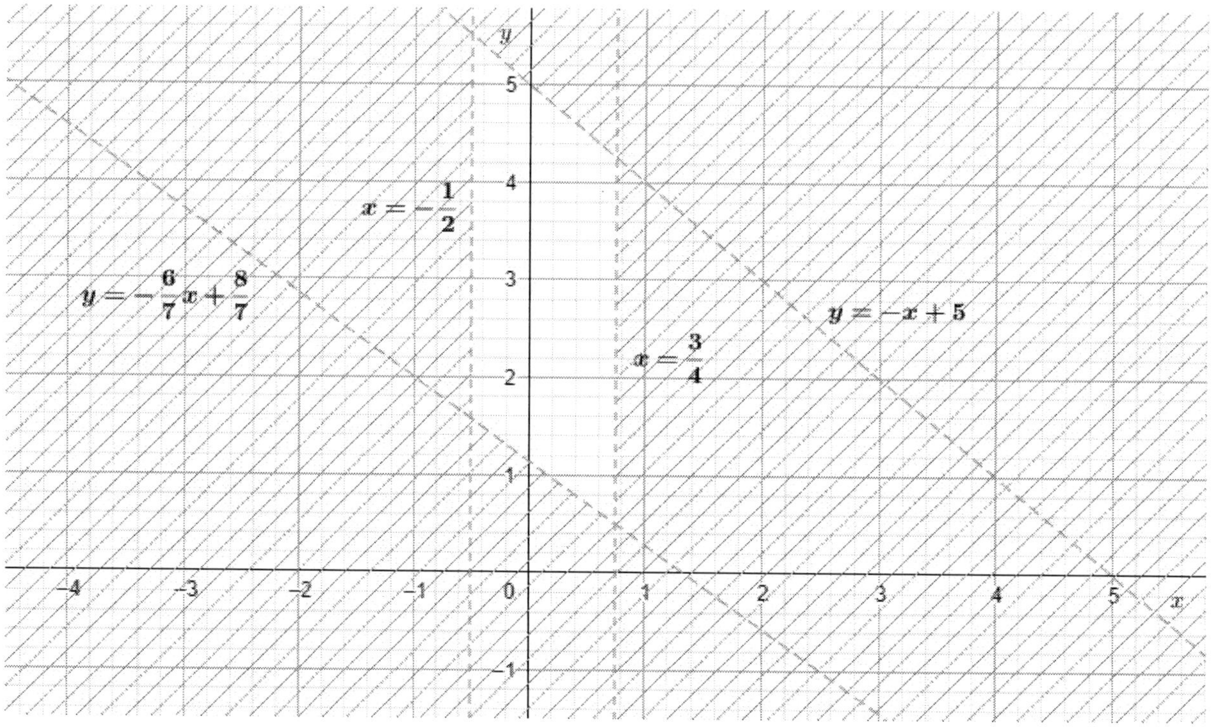


The unshaded region is the required area. This is a trapezium; substituting $x = -\frac{1}{2}$ and $x = \frac{3}{4}$ into $y = -x + 5$ and $y = -\frac{6}{7}x + \frac{8}{7}$ gives us the coordinates of the vertices:

$$\left(-\frac{1}{2}, \frac{11}{2}\right), \left(-\frac{1}{2}, \frac{11}{7}\right), \left(\frac{3}{4}, \frac{17}{4}\right), \left(\frac{3}{4}, \frac{1}{2}\right)$$

So the distance between the parallel sides is

$$\frac{3}{4} - \left(-\frac{1}{2}\right) = \frac{5}{4}$$

And the parallel sides have lengths

$$\frac{11}{2} - \frac{11}{7} = \frac{55}{14}$$

and

$$\frac{17}{4} - \frac{1}{2} = \frac{15}{4}$$

Thus the area of the trapezium is

$$\frac{1}{2} \times \left(\frac{55}{14} + \frac{15}{4}\right) \times \frac{5}{4}$$

$$= \mathbf{4.80}$$

---

57.

$$y = \frac{(x+1)(x+2)}{3x}$$

$$= \frac{x^2 + 3x + 2}{3x}$$

$$= \frac{1}{3}x + 1 + \frac{2}{3}x^{-1}$$

So at the stationary points,

$$\frac{dy}{dx} = \frac{1}{3} - \frac{2}{3}x^{-2} = 0$$

$$x^2 - 2 = 0$$

$$x^2 = 2$$

$$x = \pm\sqrt{2}$$

When $x = \sqrt{2}$,

$$y = \frac{1}{3}\sqrt{2} + 1 + \frac{2}{3\sqrt{2}}$$

$$= 1 + \frac{1}{3}\sqrt{2} + \frac{1}{3}\sqrt{2}$$

$$= 1 + \frac{2}{3}\sqrt{2}$$

To determine the nature of the turning point, note that when $x = \sqrt{2} - 0.01$,

$$\frac{dy}{dx} = \frac{1}{3} - \frac{2}{3}\left(\sqrt{2} - 0.01\right)^{-2} = -0.0047 \ldots < 0$$

and when $x = \sqrt{2} + 0.01$,

$$\frac{dy}{dx} = \frac{1}{3} - \frac{2}{3}\left(\sqrt{2} + 0.01\right)^{-2} = 0.0046 \ldots > 0$$

So it is a minimum.

When $x = -\sqrt{2}$,

$$\begin{aligned} y &= -\frac{1}{3}\sqrt{2} + 1 - \frac{2}{3\sqrt{2}} \\ &= 1 - \frac{1}{3}\sqrt{2} - \frac{1}{3}\sqrt{2} \\ &= 1 - \frac{2}{3}\sqrt{2} \end{aligned}$$

To determine the nature of this turning point, note that when $x = -\sqrt{2} - 0.01$,

$$\frac{dy}{dx} = \frac{1}{3} - \frac{2}{3}\left(-\sqrt{2} - 0.01\right)^{-2} = 0.0046 \ldots > 0$$

and when $x = -\sqrt{2} + 0.01$,

$$\frac{dy}{dx} = \frac{1}{3} - \frac{2}{3}\left(-\sqrt{2} + 0.01\right)^{-2} = -0.0047 \ldots < 0$$

So it is a maximum.

Hence there are two stationary points:

$(\sqrt{2}, 1 + \frac{2}{3}\sqrt{2})$, which is a **minimum**, and

$(-\sqrt{2}, 1 - \frac{2}{3}\sqrt{2})$, which is a **maximum**.

58. The transformation from $y = f(x)$ to $y = f(1+x)$ is a translation 1 unit to the left.

The transformation from $y = f(1+x)$ to $y = f(1-x)$ is a reflection in the $y$-axis.

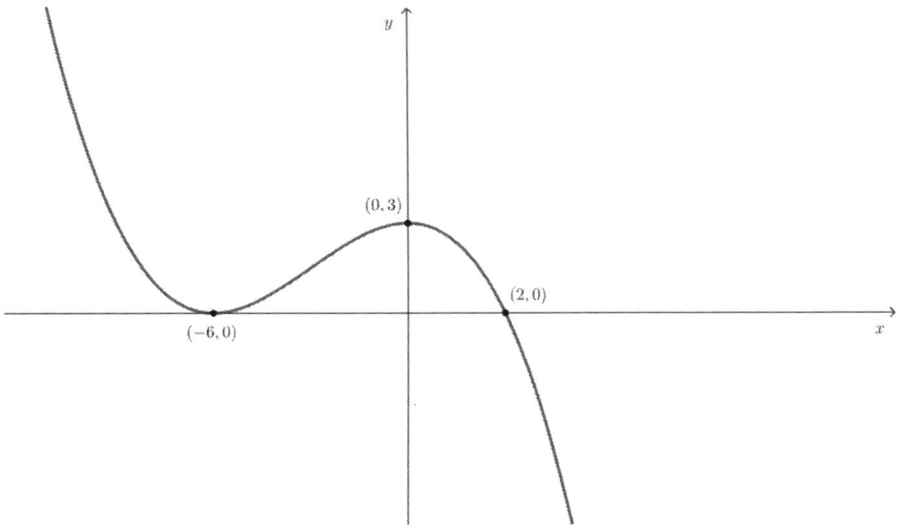


(b) The combined effect is that of a single reflection in the line $x = \frac{1}{2}$

59. The mean is $\frac{1+3+x}{3} = \frac{4+x}{3}$

If $x < 1$, then the range is $3 - x$ and the median is $1$

Then

$$3 - x = 1 + \frac{4 + x}{3}$$

$$9 - 3x = 3 + 4 + x$$

$$4x = 2$$

$$x = \frac{1}{2}$$

If $1 \leq x < 3$, then the range is $2$ and the median is $x$

Then

$$2 = x + \frac{4 + x}{3}$$

$$6 = 3x + 4 + x$$

$$4x = 2$$

$$x = \frac{1}{2}$$

But this doesn't lie in the required range; so there are no solutions between 1 and 3

Finally, if $x \geq 3$, then the range is $x - 1$ and the median is $3$

Then

$$x - 1 = 3 + \frac{4 + x}{3}$$

$$3x - 3 = 9 + 4 + x$$

$$2x = 16$$

$$x = 8$$

To check, observe that the sets $(\frac{1}{2}, 1,3)$ and $(1,3,8)$ both satisfy the given condition.

Thus $\boldsymbol{x = \frac{1}{2}}$ **or 8**

60.

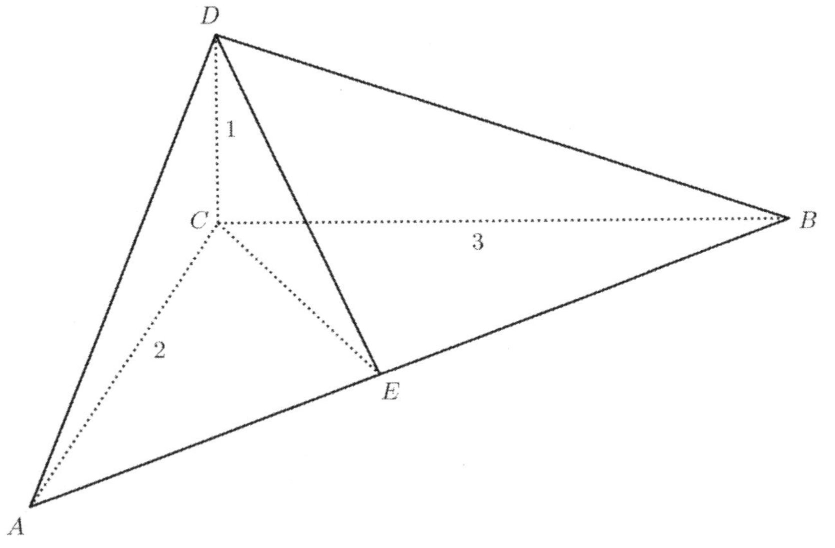


We are looking for angle $CED$, where $E$ is the point on $AB$ such that $CE$ is perpendicular to $AB$. (It happens to be the case that $DE$ is perpendicular to $AB$ as well.)

We already have $CD = 1$ in triangle $CDE$. We can find $CE$ by calculating the area of triangle $ABC$ in two different ways.

By Pythagoras,

$$AB^2 = 2^2 + 3^2 = 13$$

$$AB = \sqrt{13}$$

Now, by considering the area of triangle $ABC$ twice, we have

$$\frac{1}{2} \times 2 \times 3 = \frac{1}{2} \times \sqrt{13} \times CE$$

and so

$$CE = \frac{6}{\sqrt{13}}$$

Finally the required angle is

$$\tan^{-1}\left(\frac{CD}{CE}\right)$$

$$= \tan^{-1}\left(\frac{\sqrt{13}}{6}\right)$$

$$= \mathbf{31.0^\circ}$$

61. Since all the blocks face the same direction, we require the side length to be a multiple of 24, 42 and 99

We have

$$24 = 2^3 \times 3$$

$$42 = 2 \times 3 \times 7$$

$$99 = 3^2 \times 11$$

So the lowest common multiple of 24, 42 and 99 is

$$2^3 \times 3^2 \times 7 \times 11 = 56 \times 99 = 5544$$

This works; we can arrange the blocks with $\frac{5544}{24} = 231$ blocks running left to right, $\frac{5544}{42} = 132$ blocks running from front to back and $\frac{5544}{99} = 56$ running front top to bottom.

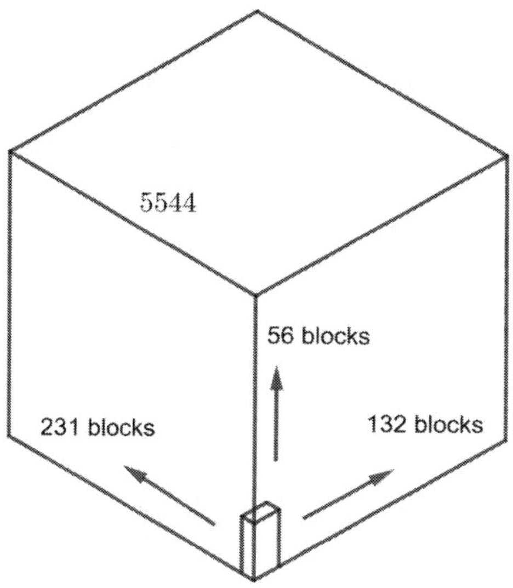


Hence the side length of the smallest possible cube is **5544**

(b) The sides of the big rectangle will be in the ratio $10:16$, so call the side lengths $10n$ and $16n$

Since we need to be able to arrange the $10 \times 16$ rectangles both vertically and horizontally, we need both $10n$ and $16n$ to be divisible by both 10 and 16

$10n$ is always a multiple of 10, and $16n$ is always a multiple of 16

Now

$$10 = 2 \times 5$$

$$16 = 2^4$$

So for $10n$ to be a multiple of 16, we require $n$ to be a multiple of $2^3 = 8$, and for $16n$ to be a multiple of 10, we require $n$ to be a multiple of 5

Thus the smallest possible value of $n$ is the lowest common multiple of 8 and 5, which is $8 \times 5 = 40$

This would lead to rectangle with dimensions 400 and 640. In fact, this works; vertically, we can arrange 64 little rectangles from left to right, and 25 small rectangles from top to bottom.

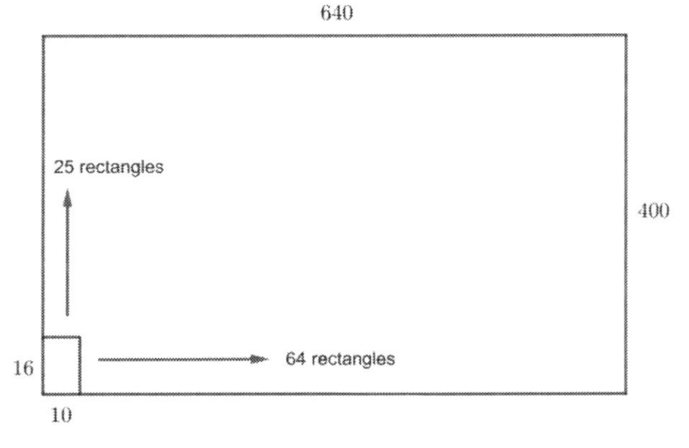


Horizontally, we can arrange 40 little rectangles from left to right, and 40 rectangles from top to bottom.

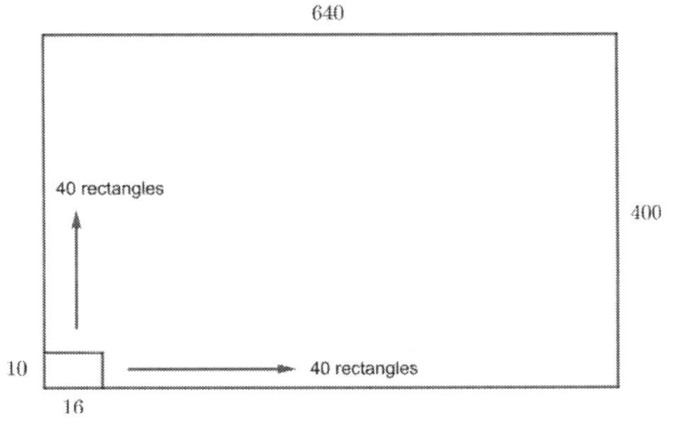


Hence the sides of the smallest possible rectangle with all three properties are **400 and 640**

62. Suppose an odd number of blue marbles go into the green bag. Then an even number of yellow marbles go into the green bag. This can be done in $25 \times 25$ ways.

If instead an even number of blue marbles go into the green bag, then an odd number of yellow marbles go into the green bag. This can be done in $26 \times 25$ ways.

But we must remove the case where all 50 blue marbles and all 49 yellow marbles end up in the green bag (since then the marbles haven't been split into two groups).

Thus the total number of ways to split the marbles is

$$25 \times 25 + 26 \times 25 - 1 = \mathbf{1274}$$

---

63. If $x > -1$, then

$$x > 2(x+1)$$

$$x > 2x + 2$$

$$x < -2$$

But this is impossible as $x > -1$

If $x = -1$ then the left hand side involves division by zero, which isn't allowed.

So $x < -1$, and we have

$$x < 2(x+1)$$

$$x < 2x + 2$$

$$x > -2$$

Hence the final answer is

$$\mathbf{-2 < x < -1}$$

---

64.

$$9^{8^7} = \left(3^2\right)^{8^7} = 3^{2\times 8^7} = 3^{2\times\left(2^3\right)^7} = 3^{2\times 2^{21}} = 3^{2^{22}} = 3^{\left(2^2\right)^{11}} = 3^{4^{11}}$$

Hence $n = \mathbf{11}$

65.

$$\frac{\sqrt[3]{4x}}{\sqrt[3]{15}} + 2 = \frac{\sqrt[3]{9x}}{\sqrt[3]{10}}$$

$$\sqrt[3]{10}\sqrt[3]{4x} + 2\sqrt[3]{10}\sqrt[3]{15} = \sqrt[3]{15}\sqrt[3]{9x}$$

$$\sqrt[3]{15}\sqrt[3]{9x} - \sqrt[3]{10}\sqrt[3]{4x} = 2\sqrt[3]{10}\sqrt[3]{15}$$

$$\sqrt[3]{x}\left(\sqrt[3]{15}\sqrt[3]{9} - \sqrt[3]{10}\sqrt[3]{4}\right) = 2\sqrt[3]{150}$$

$$\sqrt[3]{x}\left(\sqrt[3]{135} - \sqrt[3]{40}\right) = 2\sqrt[3]{150}$$

$$\sqrt[3]{x}\left(\sqrt[3]{27}\sqrt[3]{5} - \sqrt[3]{8}\sqrt[3]{5}\right) = 2\sqrt[3]{150}$$

$$\sqrt[3]{x}\left(3\sqrt[3]{5} - 2\sqrt[3]{5}\right) = 2\sqrt[3]{150}$$

$$\sqrt[3]{x}\sqrt[3]{5} = 2\sqrt[3]{150}$$

$$\sqrt[3]{x} = 2\sqrt[3]{30}$$

$$x = 2^3 \times 30$$

$$= \mathbf{240}$$

66.

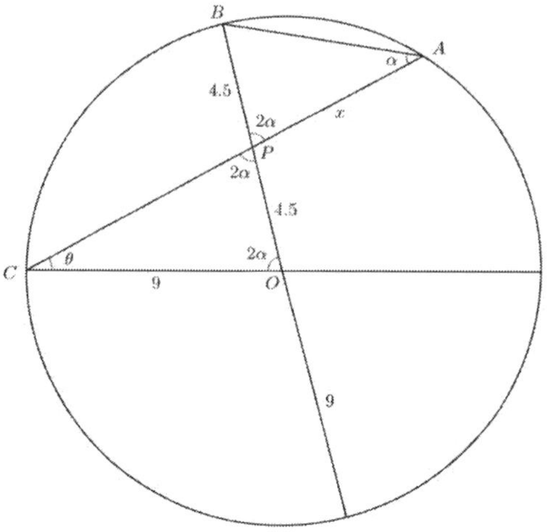


(a) Let $PAB = \alpha$ . Then $APB = 2\alpha$ and so $OPC = 2\alpha$ (vertically opposite angles).

Also, $COP = 2 \times PAB = 2\alpha$ (angle at the centre is twice angle at the circumference).

Hence triangle $COP$ is isosceles and so $CP = 9$

Since $P$ is the midpoint of $OB$ , we have $OP = 4.5$

By the cosine rule,

$$\cos\theta = \frac{9^2 + 9^2 - 4.5^2}{2 \times 9 \times 9}$$

$$= \frac{7}{8}$$

Thus

$$\theta = \cos^{-1}\frac{7}{8}$$

$$= \mathbf{28.96°}$$

(b) Extend radius $BO$ until it reaches the other side of the whole circle. Now considering chords intersecting at $P$ , we have

$$9x = 4.5 \times 13.5$$

(by the Intersecting Chords Theorem),

and so

$$x = \mathbf{6.75}$$

67.

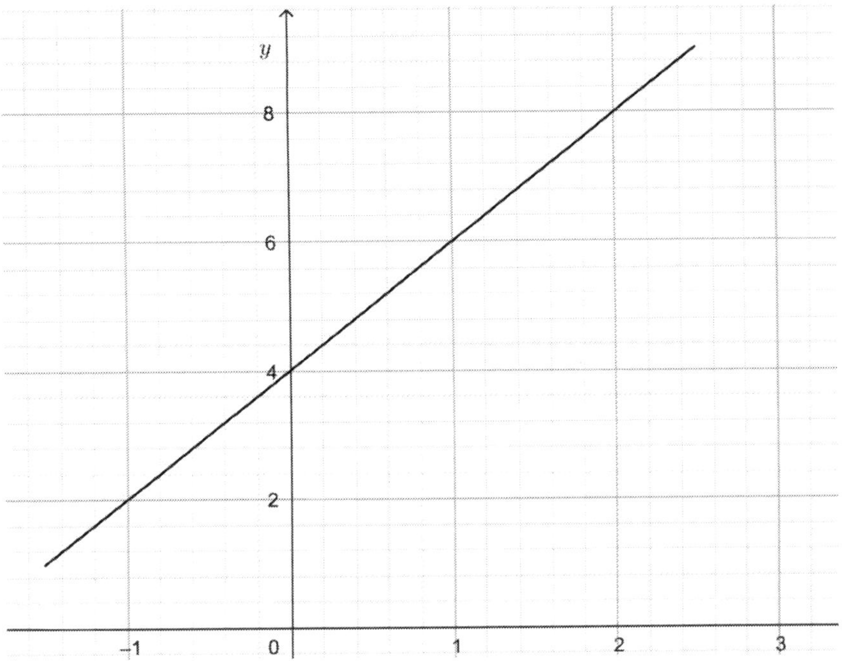


The line above has gradient 2 and $y$-intercept 4, so its equation is

$$y = 2x + 4$$

$P$ is a point on the line, so $q = 2p + 4$

It follows that $\begin{pmatrix} p+2 \\ q \end{pmatrix} = \begin{pmatrix} p+2 \\ 2p+4 \end{pmatrix} = (p+2)\begin{pmatrix} 1 \\ 2 \end{pmatrix}$ and $\begin{pmatrix} p-3 \\ q-10 \end{pmatrix} = \begin{pmatrix} p-3 \\ 2p-6 \end{pmatrix} = (p-3)\begin{pmatrix} 1 \\ 2 \end{pmatrix}$

So both vectors are parallel to $\begin{pmatrix} 1 \\ 2 \end{pmatrix}$

Finally, considering the graph again we see that $0 < q < 10$; in particular, the $q$ in $\begin{pmatrix} p+2 \\ q \end{pmatrix}$ is positive but the $q - 10$ in $\begin{pmatrix} p-3 \\ q-10 \end{pmatrix}$ is negative.

This, combined with the fact that the vectors are parallel, proves that the vectors point in opposite directions.

68. While $x = 30°$ is a solution (as then $\sin 2x = \sin 60° = \frac{\sqrt{3}}{2}$ ), the graph used by the student ( $y = \sin x$ ) shows us that

$$\sin 30° = \sin 150°$$

and this fact isn't helpful, as both of these are equal to $\frac{1}{2}$, not $\frac{\sqrt{3}}{2}$

Once the first solution $x = 30°$ was found, the student would have had better luck using the graph $y = \sin 2x$ instead.

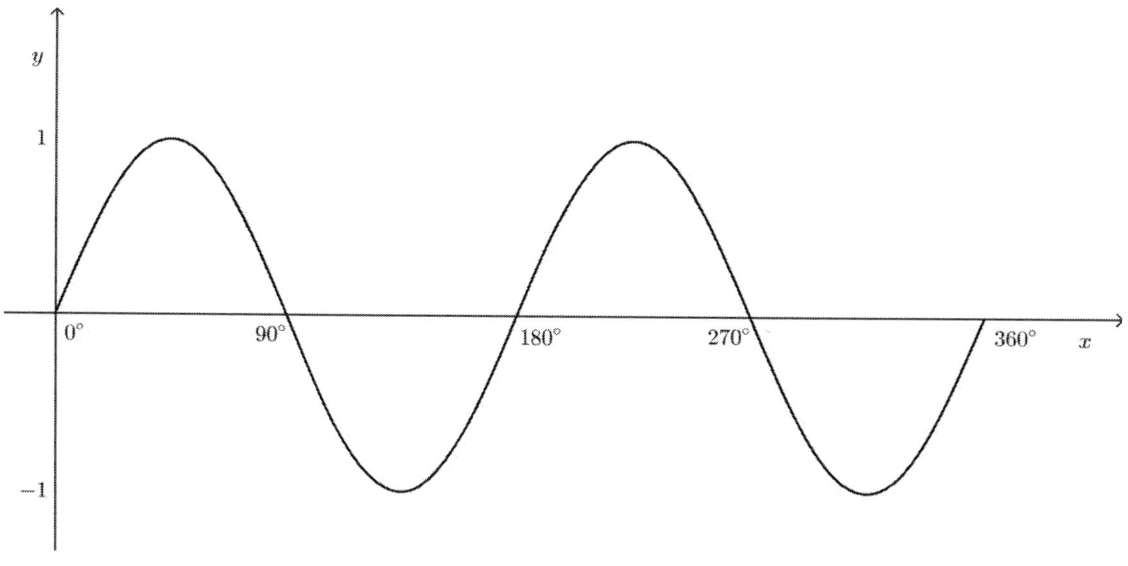


Now we see that that, since $x = 30°$ is a solution, so are $x = 60°, 210°$ and $240°$

So the complete set of solutions is

$$x = \mathbf{30°, 60°, 210°, 240°}$$

69.

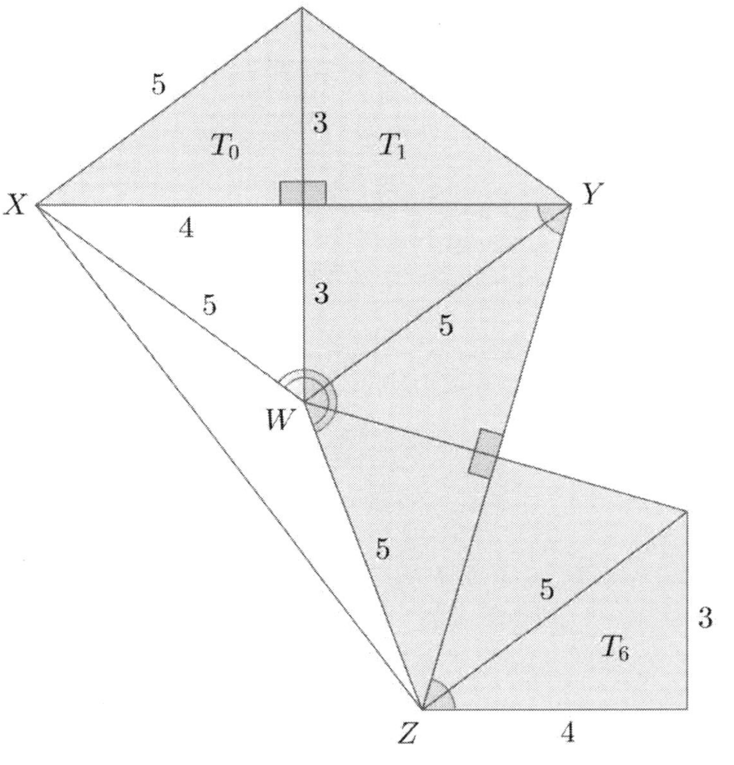


(a) Since $3^2 + 4^2 = 5^2$, triangle $T_0$ is right-angled, and so $XY$ and $YZ$ are straight lines.

For both $T_0$ and $T_6$, the sides are 3, 4 and 5 in clockwise order, so no reflection is required to transform one into the other.

Finally, the four angles marked with a single arc are all equal; hence by alternate angles, the base of $T_0$ is parallel to the base of $T_6$. So no rotation is required either.

Thus a single **translation** will transform $T_0$ into $T_6$

(b) The lower left vertex of the wooden triangle moves directly from $X$ to $Z$ as it slides, so we want the length of $XZ$

Observing the sides of the triangles, we find that each of the angles marked with a double arc is equal to

$$\tan^{-1}\left(\frac{4}{3}\right)$$

So

$$\angle ZWX = 360° - 4\tan^{-1}\left(\frac{4}{3}\right)$$

$$= 147.47\ldots°$$

Finally, by the cosine rule we have

$$XZ^2 = 5^2 + 5^2 - 2(5)(5)\cos(147.47\ldots°)$$

$$= \frac{2304}{25}$$

And so the distance moved by the wooden triangle is

$$XZ = \sqrt{\frac{2304}{25}}$$

$$= \mathbf{9.6}$$

70. Extend the diagram to form a cone.

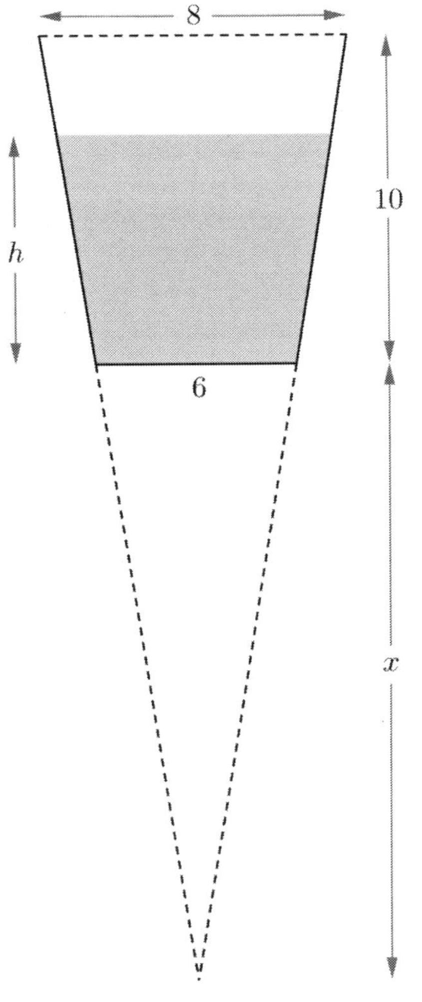


We now have three mathematically similar cones:

- a small cone with height $x$
- a medium cone with height $x + h$
- a large cone with height $x + 10$

As the small and large cones are similar, comparing base diameters and heights gives

$$\frac{x+10}{8} = \frac{x}{6}$$

$$3(x+10) = 4x$$

$$3x + 30 = 4x$$

$$x = 30$$

So the volume of the small cone is

$$\frac{1}{3} \times \pi \times 3^2 \times 30 = 90\pi$$

The volume of liquid is 250cm$^3$

As the small and middle cones are similar, comparing volumes and heights gives

$$\left(\frac{30+h}{30}\right)^3 = \frac{90\pi + 250}{90\pi} = 1.8841 \ldots$$

So

$$\frac{30+h}{30} = \sqrt[3]{1.8841 \ldots} = 1.2351 \ldots$$

$$30 + h = 37.053 \ldots$$

$$h = 7.0535 \ldots$$

Hence the height of the surface of the liquid above the base of the beaker is

**7.05 cm**

71. Let £$x$ be their pre-tax salaries.

If $x \leq 10000$ then neither Emmy nor Sophie is taxed, and so their post-tax salaries will be equal.

If $10001 \leq x \leq 15000$, then Emmy is taxed but Sophie isn't, so their post-tax salaries cannot possibly be equal.

If $15001 \leq x \leq 50000$, then Emmy's post-tax salary is

$$10000 + 0.8(x - 10000) = 0.8x + 2000$$

and Sophie's post-tax salary is

$$15000 + 0.7(x - 15000) = 0.7x + 4500$$

So we have

$$0.8x + 2000 = 0.7x + 4500$$

$$0.1x = 2500$$

$$x = 25000$$

Finally, if $x \geq 50001$, then Emmy's post-tax salary is

$$10000 + 0.8(40000) + 0.5(x - 50000) = 0.5x + 17000$$

and Sophie's post-tax salary is still

$$0.7x + 4500$$

So we have

$$0.5x + 17000 = 0.7x + 4500$$

$$0.2x = 12500$$

$$x = 62500$$

Thus the set of possible pre-tax salaries is

$$\mathbf{£1, £2, £3, \ldots, £9999, £10000, £25000, £62500}$$

72. (a) Consider the diagram below

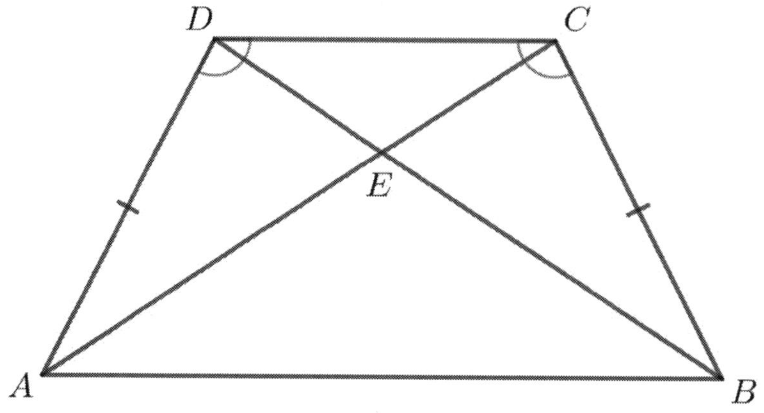


(i) For triangles $ACD$ and $BDC$ we have

$AD = BC$ (given)

$ADC = BCD$ (given)

$DC = CD$ (common side)

Hence by SAS, triangles $ACD$ and $BDC$ are congruent, and so the diagonals $AC$ and $BD$ are equal in length.

(ii) As triangles $ACD$ and $BDC$ are congruent, we have

$$ACD = BDC$$

It follows that triangle $DEC$ is isosceles, and so $DE = CE$

But from (i) we also have $AC = BD$; hence $AE = BE$

Now consider triangles $AEB$ and $DEC$. We have

$\frac{AE}{DE} = \frac{BE}{CE}$

$AEB = DEC$ (vertically opposite angles)

So by SAS, triangles $AEB$ and $DEC$ are similar and isosceles. Thus $BAC = ACD$, and so by alternate angles the two unmarked sides are parallel.

(b)

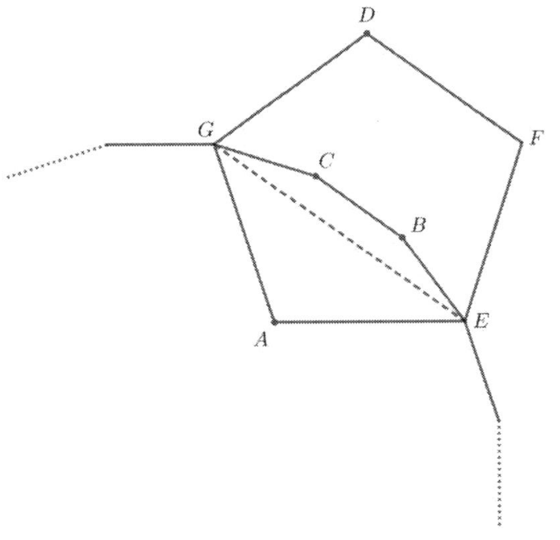


Each interior angle of a regular pentagon is equal to

$$\frac{180°(5-2)}{5} = 108°$$

Each interior angle of a regular icosagon is equal to

$$\frac{180°(20-2)}{20} = 162°$$

So $GAE = 108°$, and since triangle $GAE$ is isosceles, we also have

$AGE = AEG = \frac{180° - 108°}{2} = 36°$

Also, $GCB = CBE = 162°$ and $GC = BE$, so by the result of (a)(ii), the line segments $GE$ and $CB$ are parallel.

Hence $CGE = BEG = 180° - 162° = 18°$

Importantly, we now have that $AEB = AGC = 36° + 18° = 54°$, as well as
$CGD = 108° - 54° = 54°$

Finally, consider triangles $AEB$ and $DGC$; we have

$AE = DG$ (sides of a regular pentagon)

$AEB = DGC = 54°$

$EB = GC$ (sides of a regular icosagon)

Hence by SAS triangles $AEB$ and $DGC$ are congruent, and so the distances $AB$ and $CD$ are equal as required.

73. Let $x$ be the number of carrots that Carlotta ends up chopping. Then once the preparation is over, Carlotta would have peeled 10 potatoes and chopped $x$ carrots, and her son would have chopped $24-x$ carrots.

Carlotta would have worked for

$$10p+qx$$

seconds in total.

Her son would have worked for

$$r(24-x)$$

seconds in total.

Since they finish at the same time, we have

$$10p+qx=r(24-x)$$

$$10p+qx=24r-rx$$

$$qx+rx=24r-10p$$

$$x(q+r)=24r-10p$$

$$x=\frac{24r-10p}{q+r}$$

Hence the total preparation time is

$$10p+qx$$

$$=10p+\frac{q(24r-10p)}{q+r}$$

$$=\frac{10p(q+r)}{q+r}+\frac{q(24r-10p)}{q+r}$$

$$=\frac{10p(q+r)+q(24r-10p)}{q+r}$$

$$=\frac{10pq+10pr+24qr-10pq}{q+r}$$

$$=\frac{\mathbf{10pr+24qr}}{\mathbf{q+r}}$$

74.

(i)

$$\overrightarrow{AB} = \overrightarrow{AO} + \overrightarrow{OB}$$

$$= -\boldsymbol{a} + \boldsymbol{b}$$

(ii)

$$\overrightarrow{OP} = \overrightarrow{OA} + \frac{1}{3}\overrightarrow{AB}$$

$$= \boldsymbol{a} + \frac{1}{3}(-\boldsymbol{a} + \boldsymbol{b})$$

$$= \frac{2}{3}\boldsymbol{a} + \frac{1}{3}\boldsymbol{b}$$

(iii)

$$\overrightarrow{BQ} = \overrightarrow{BO} + \overrightarrow{OQ}$$

$$= -\boldsymbol{b} + \frac{1}{2}\boldsymbol{a}$$

$$= \frac{1}{2}\boldsymbol{a} - \boldsymbol{b}$$

(iv)

Since $O, R$ and $P$ are collinear, we can write

$$\overrightarrow{OR} = x\overrightarrow{OP}$$

$$= x(\frac{2}{3}\boldsymbol{a} + \frac{1}{3}\boldsymbol{b})$$

$$= \frac{2x}{3}\boldsymbol{a} + \frac{x}{3}\boldsymbol{b}$$

Similarly, since $B, R$ and $Q$ are collinear, we can write

$$\overrightarrow{BR} = y\overrightarrow{BQ}$$

$$\overrightarrow{BR} = y(-\boldsymbol{b} + \frac{1}{2}\boldsymbol{a})$$

It follows that

$$\overrightarrow{OR} = \overrightarrow{OB} + \overrightarrow{BR}$$

$$= \boldsymbol{b} + y(-\boldsymbol{b} + \frac{1}{2}\boldsymbol{a})$$

$$= \frac{y}{2}\boldsymbol{a} + (1 - y)\boldsymbol{b}$$

We now equate the two expressions for $\overrightarrow{OR}$:

$$\frac{2x}{3}\boldsymbol{a} + \frac{x}{3}\boldsymbol{b} = \frac{y}{2}\boldsymbol{a} + (1 - y)\boldsymbol{b}$$

Since $\boldsymbol{a}$ and $\boldsymbol{b}$ are not parallel, the corresponding parts in the equation can be equated:

$$\frac{2x}{3} = \frac{y}{2}$$

$$\frac{x}{3} = 1 - y$$

We can now solve simultaneously to find $x$ and $y$:

$$y = \frac{4x}{3}$$

So

$$\frac{x}{3} = 1 - \frac{4x}{3}$$

$$x = 3 - 4x$$

$$5x = 3$$

$$x = \frac{3}{5}$$

Hence

$$y = \frac{4}{3} \times \frac{3}{5} = \frac{4}{5}$$

Finally, we have

$$\overrightarrow{OR} = \frac{2x}{3}\boldsymbol{a} + \frac{x}{3}\boldsymbol{b}$$

$$= \frac{2}{5}\boldsymbol{a} + \frac{1}{5}\boldsymbol{b}$$

75. (a)

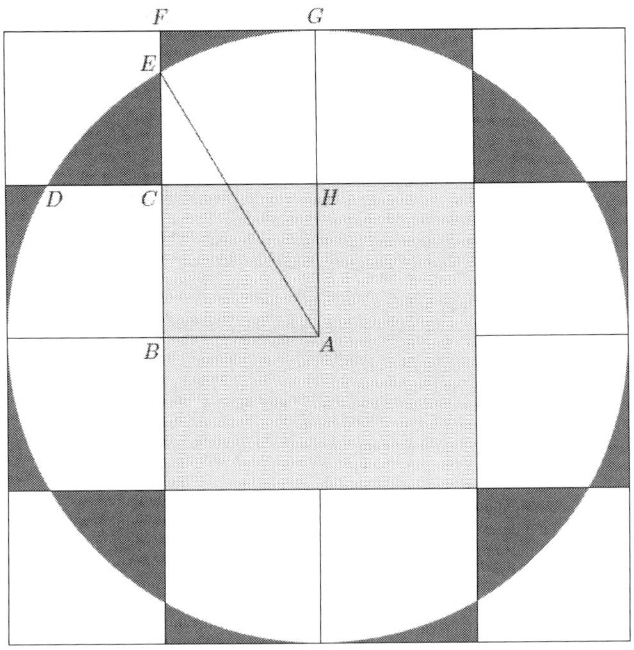


Add lines to the diagram as shown.

Consider triangle $ABE$

$AB = 1$

$AE = 2$ (radius)

$ABE = 90°$

So $BE = \sqrt{2^2 - 1^2} = \sqrt{3}$ (Pythagoras)

And hence the area of $ABE$ is

$$\frac{1}{2} \times 1 \times \sqrt{3} = \frac{\sqrt{3}}{2}$$

We also have

$$\cos BAE = \frac{1}{2}$$

and so

$$BAE = 60°$$

and

$$EAG = 30°$$

This means that sector $EAG$ represent one twelfth on the entire circle, so its area is

$$\frac{1}{12} \times \pi \times 2^2 = \frac{\pi}{3}$$

The grey region $EFG$ combines with the sector $EAG$ and triangle $ABE$ to cover exactly two paving stones.

So

$$\text{Area } (EFG) = 2 - \frac{\pi}{3} - \frac{\sqrt{3}}{2}$$

and then

$$\text{Area}(CEGH) = 1 - \text{area}(EFG) = \frac{\pi}{3} + \frac{\sqrt{3}}{2} - 1$$

Next, the grey region $CDE$ combines with two white regions like $CEGH$ and a single square of side 1 to cover a quarter of the circle.

So

$$\text{area}(CDE) = \frac{1}{4} \times \pi \times 2^2 - 1 - 2 \times \text{area}(CEGH) = 1 + \frac{\pi}{3} - \sqrt{3}$$

Finally, the total grey area comprises 8 regions like $EFG$ and 4 regions like $CDE$

Thus the final answer is

$$8\left(2 - \frac{\pi}{3} - \frac{\sqrt{3}}{2}\right) + 4(1 + \frac{\pi}{3} - \sqrt{3})$$

$$= \mathbf{20 - \frac{4\pi}{3} - 8\sqrt{3}\ m^2}$$

(b)

Imagine drawing a vertical line down the centre of the diagram. The design is identical on either side of this line, except that any grey regions on the left are white on the right, and vice versa.

Hence exactly half of the paving in grey, and so the total grey paved area is

**6 m²**

# APPENDIX

## Alternative Solutions

For each problem in the book, one solution has been provided. However, this does not preclude other possible solutions. If you have found another, possibly more elegant way of solving any of the problems, feel free to let me know at

elasticnumbers@gmail.com

## To find out more...

If you have any questions or comments, you can contact me at the email address above.

For discussions on some fantastic problems from around the world, visit my blog at

puzzlecritic.wordpress.com

You can also find me on Twitter (@puzzlecritic).

## Also by Daniel Griller

# ELASTIC NUMBERS: 108 PUZZLES FOR THE SERIOUS PROBLEM SOLVER

"Such a pleasure. A unique and wonderful book!"

Professor Imre Leader, University of Cambridge

*Ashley starts listing numbers 1, 2, 3, 4, 6, 8, 9, 10, ..., missing out every number containing a digit 5 or 7. What is the 1000th number in his list?*

**This is not an ordinary puzzle book.** Beautifully crafted and immensely enjoyable, the problems require minimal technical knowledge, being accessible to young secondary school pupils. However, there is an astonishing range in difficulty; while some of the problems are fairly straightforward, others are significantly tougher, with a great deal of ingenuity and clarity of thought needed to make progress.

Whether you are a student preparing for a maths competition, an educational establishment seeking to supplement your problem solving resources, or an individual looking for a different sort of challenge, *Elastic Numbers* is a unique collection, and will push you to the very edge of your abilities.

Never stop learning

Printed in Great Britain
by Amazon

13941186R00088